SpringerBriefs in Probability and Mathematical Statistics

SpringerBriefs present concise summaries of cutting-edge research and practical applications across a wide spectrum of fields. Featuring compact volumes of 50 to 125 pages, the series covers a range of content from professional to academic. Briefs are characterized by fast, global electronic dissemination, standard publishing contracts, standardized manuscript preparation and formatting guidelines, and expedited production schedules.

Typical topics might include:

- A timely report of state-of-the art techniques
- A bridge between new research results, as published in journal articles, and a contextual literature review
- A snapshot of a hot or emerging topic
- Lecture of seminar notes making a specialist topic accessible for non-specialist readers
- SpringerBriefs in Probability and Mathematical Statistics showcase topics of current relevance in the field of probability and mathematical statistics

Manuscripts presenting new results in a classical field, new field, or an emerging topic, or bridges between new results and already published works, are encouraged. This series is intended for mathematicians and other scientists with interest in probability and mathematical statistics. All volumes published in this series undergo a thorough refereeing process.

The SBPMS series is published under the auspices of the Bernoulli Society for Mathematical Statistics and Probability.

All titles in this series are peer-reviewed to the usual standards of mathematics and its applications.

More information about this series at http://www.springer.com/series/14353

Zheng Gao · Stilian Stoev

Concentration of Maxima and Fundamental Limits in High-Dimensional Testing and Inference

Bernoulli Society
for Mathematical Statistics
and Probability

Zheng Gao (iD)
Department of Statistics
University of Michigan–Ann Arbor
Ann Arbor, MI, USA

Stilian Stoev (iD)
Department of Statistics
University of Michigan–Ann Arbor
Ann Arbor, MI, USA

ISSN 2365-4333 ISSN 2365-4341 (electronic)
SpringerBriefs in Probability and Mathematical Statistics
ISBN 978-3-030-80963-8 ISBN 978-3-030-80964-5 (eBook)
https://doi.org/10.1007/978-3-030-80964-5

This Springer imprint is published by the registered company Springer Nature Switzerland AG
The registered company address is: Gewerbestrasse 11, 6330 Cham, Switzerland

To our teachers and families with deepest gratitude

Preface

This text presents a collection of new results and recent developments on the phase-transition phenomena in sparse signal problems. The main theme is the study of the fundamental limits in high-dimensional testing and inference. Since the seminal works of Ingster (1998) and Donoho and Jin (2004), the subject has received a lot of attention in the literature with important contributions from Ji and Jin (2012); Genovese et al. (2012); Jin et al. (2014); Arias-Castro and Chen (2017); Butucea et al. (2018). These works, among many others, have discovered some fundamental limits in the so-called *needle in haystack* problems, where a sparse signal is observed with high-dimensional additive noise. In this setting, two archetypal problems arise—the *signal detection* and *signal support recovery*. The signal detection refers to a global hypothesis testing problem that amounts to determining the presence of non-zero signal in any of its dimensions. The support recovery, on the other hand, can be seen either as a multiple testing problem where the presence of non-zero signal is tested for each signal location of interest, or alternatively, as an inference problem that aims to estimate the signal support, i.e., the locations of the non-zero signal components. The fundamental limits of these problems are studied in the so-called high-dimensional asymptotic regime where the dimension p of the underlying signal grows to infinity, and the sample size n is either bounded or grows slowly relative to p.

From a probabilistic perspective, these aforementioned fundamental limits are stated as asymptotic zero-one type laws, as dimensionality diverges. Namely, consider a sparse signal with *support size* on the order of $p^{1-\beta}$ for some parameter $\beta \in (0, 1)$. Parameterize the non-zero *signal amplitude* by $\Delta(p^r)$, for some $r > 0$ and a suitable monotone non-decreasing function $\Delta(\cdot)$. Then, for a broad range of error distributions and statistical problems, one encounters a sharp transition between the regimes where the problem is solvable and unsolvable depending on the signal magnitude r and signal sparsity β. More precisely, there exists a boundary function $f(\beta)$ such that if the signal magnitudes are *above* the boundary, $r > f(\beta)$, then the problem can be solved with vanishing loss as $p \to \infty$, with a suitable statistical procedure. On the other hand, if the signal is below that same boundary, i.e., $r < f(\beta)$, all statistical procedures fail to provide a solution with a vanishing loss, as $p \to \infty$. Of course, depending on whether one considers the detection (testing) or support recovery (inference) problems, different loss functions quantify success and failure. The choice of the loss functions is often guided by the applications, resulting in a rich picture of phase transitions (see, e.g., Fig. 3.2).

The contributions of this work. The fundamental limits of the classic detection problem hinge of the analysis of the discrepancy between the *null* and *alternative* hypotheses, e.g., via Hellinger distance. Thus, perhaps for technical reasons, much of the analysis in the existing literature has been done under the assumption that the additive errors are independent and/or Gaussian, or using loss functions unaffected by the dependence such as the Hamming loss. In this work, we demonstrate that the support recovery problems, especially *exact support recovery*, are best understood from the novel perspective of the *concentration of maxima* phenomenon in extreme value theory. It turns out that under a very broad range of light-tailed error distributions and under a *very* broad range of error-dependence structures, the maxima of the errors, when rescaled (but not centered!) converge in probability to a positive constant. This concentration property leads to a complete solution of the exact support recovery problem for the broad family of thresholding procedures. Most, if not all, existing support estimation procedures are types of thresholding procedures (see Sect. 2.2). That is, the signal support estimate comprises all components exceeding a suitable (potentially data-dependent) threshold. We show, by exploiting concentration of maxima, that thresholding procedures obey a phase transition, where if the signal is above a certain boundary, asymptotically exact recovery is possible while below the boundary all thresholding procedures fail, as $p \to \infty$. Remarkably, light-tailed maxima concentrate under very broad and strong dependence. This is exemplified by our characterization of the concentration of maxima phenomenon for Gaussian triangular arrays. For example, in the special case of stationary Gaussian time series, vanishing autocovariance is necessary and sufficient for the maxima to concentrate in the same way as independent standard normal random variables. This is in stark contrast with the behavior of sums, commonly studied under short- and long-range dependence conditions (see, e.g., Dedecker et al. 2007; Pipiras and Taqqu 2017). Simply put, the notion of weak dependence that entails that the maxima of dependent variables concentrate at the same rate as in the case of independence is fundamentally weaker than the conventional mixing conditions widely used in the study of sums.

Our probabilistic contributions may be of independent interest and extend classic work of Berman (1964). Concentration of maxima is a type of superconcentration phenomenon studied also in Chatterjee (2014) and Tanguy (2015a). The robustness of the concentration of maxima phenomenon to dependence can perhaps explain the universality of phase transitions in support recovery problems.

The use of concentration of maxima phenomenon highlights one core idea in our work, which allows for a first of its kind comprehensive treatment of thresholding procedures under very broad error-dependence conditions. The text involves also a full spectrum of related results such as minimax-optimality and finite-sample Bayes optimality in support estimation. Using different type of loss functions and Type I error controls, we obtain a rich picture of the exact and approximate support recovery problems in high dimensions. Many of these phase-transition results have not appeared in previously published literature.

High-dimensional support recovery problems arise in many modern applications ranging from cybersecurity, theoretical computer science, to statistical genetics. Genome-wide Association Studies (GWAS) in genetics are particularly natural applications, where the asymptotic phase-transition results help explain and quantify a previously observed empirical phenomenon of the so-called *steep part of the power curve*. In the last chapter of this work, we detail this application and highlight future theoretical and practical consequences of our work.

Target audience. The original research presented in this text originates from the doctoral dissertation of the first author in the Statistics Department at the University of Michigan, Ann Arbor. The main goal of this text is to provide a comprehensive treatment of the exact and approximate support recovery problems by utilizing existing and newly developed probabilistic tools on concentration of maxima. The text also provides a quick introduction to the state of the art in the dynamic area of phase transitions in high-dimensional testing and inference. It is accessible to doctoral students in Statistics with background in measure-theoretic probability and statistics as well as to researchers in applied fields working with high-dimensional datasets. The text can be used as a reference and a supplement to a special topics course on high-dimensional inference.

Chicago, USA
Ann Arbor, USA
June 2021

Zheng Gao
Stilian Stoev

Acknowledgements The authors gratefully acknowledge the support of their families and all colleagues from the Statistics Department at the University of Michigan, Ann Arbor. Special thanks (in alphabetical order) go to Xuming He, Tailen Hsing, Michalis Kallitsis, Liza Levina, Yuanzhi Li, Rodderick Little, Ya'acov Ritov, Kerby Shedden, Jinqi Shen, Jonathan Terhorst, and Gongjun Xu. We would also like to thank the editorial board at the SpringerBriefs series in Probability and Mathematical Statistics, and Associate Editor Richard Kruel, who had been extremely supportive throughout the preparation of the manuscript. The authors were partially supported by the NSF program *Algorithms for Threat Detection*.

Contents

Acronyms

AGG	Asymptotically generalized Gaussian
BH	Benjamini–Hochberg
CDF	Cumulative distribution function
FDR	False discovery rate
FNR	False non-discovery rate
FWER	Family-wise error rate
FWNR	Family-wise non-discovery rate
GWAS	Genome-wide association studies
HC	Higher criticism
iid	Independent and identically distributed
LR	Likelihood ratio
RS	Relatively stable/relative stability
SNP	Single-nucleotide polymorphisms
URS	Uniform relatively stable/uniform relative stability

Chapter 1
Introduction and Guiding Examples

The proliferation of information technology has enabled us to collect and consume huge volumes of data at unprecedented speeds and at very low costs. This convenient access to data gave rise to a fundamentally different way of pursuing scientific questions. In contrast with the traditional hypothesis–experiment–analysis cycle where data are collected from the experiments, nowadays abundant data are often available before specific questions are even formulated. Such data can be used for not just evaluating hypotheses, but also for *generating* and *selecting* the hypotheses to pursue. As a result, multiple testing—where a large number of hypotheses are formulated and screened for their plausibility simultaneously—has become a staple of modern data-driven studies.

An archetypal example of multiple testing problems is genetic association studies (Bush and Moore 2012). In these studies, scientists test hypotheses relating each of the hundreds of thousands of genetic marker locations to phenotypic traits of interest. For a phenotypic trait on which we have little prior knowledge, we cannot simply test for association on one or a few specific genetic locations, as there are often not enough empirical evidence or biological theory to pin point these genetic locations in the first place. Rather, the goal here is to select the set of most promising genetic markers from a large number of candidate locations for subsequent investigation.

Another example of multiple testing problems arise in cybersecurity, where millions of IP addresses are monitored in real time. In this engineering application, statistics are collected and tests are performed for each IP address, in an attempt to locate the IP addresses with anomalous network activities, so that malicious traffic and volumetric attacks can be filtered to protect end users of network services (Kallitsis et al. 2016). Similar to the genetic application above, we use data to search over candidate IP addresses and identify locations of interest.

© The Author(s), under exclusive license to Springer Nature Switzerland AG 2021
Z. Gao and S. Stoev, *Concentration of Maxima and Fundamental
Limits in High-Dimensional Testing and Inference*,
SpringerBriefs in Probability and Mathematical Statistics,
https://doi.org/10.1007/978-3-030-80964-5_1

We are motivated very much by these examples to study high-dimensional multiple testing problems where a large number of hypotheses are tested simultaneously. In the rest of the introduction, we shall more review the main objectives of high-dimensional multiple testing, and elaborate on these objectives with two classes of data models in the context of various applications.

1.1 The Additive Error Model

Consider the canonical signal-plus-noise model where the observation x is a high-dimensional vector in $\mathbb{R}^p$,

$$x(i) = \mu(i) + \epsilon(i), \quad i = 1, \ldots, p. \tag{1.1}$$

The signal, $\mu = (\mu(i))_{i=1}^p$, is a vector with s non-zero components supported on the set $S = \{i : \mu(i) \neq 0\}$; the second term ϵ is a random error vector. The goal of high-dimensional statistics is usually twofold:

I. *Signal detection*: To detect the presence of non-zero components in μ. That is, to test the global hypothesis $\mu = 0$.
II. *Support recovery*: To estimate the support set S. This is also sometimes referred to as the *support estimation* or *signal identification* problem.

To illustrate, in the engineering application of cybersecurity, Internet service providers (ISP) routinely monitor a large number of network traffic streams to determine if there are abnormal surges, blackouts, or other types of anomalies. The data vector x could represent, for example, incoming traffic volumes to each server node, Internet protocol (IP) address, or port that the ISP monitors. In this case, the vector μ represents the average traffic volumes in each of the streams under normal operating conditions, and ϵ's—the fluctuations around these normal levels of traffic. The signal detection problem in this context is then equivalent to determining if there are *any* anomalies among all data streams, and the support recovery problem is equivalent to *identifying the streams experiencing anomalies*. Similar questions of signal detection and support recovery are pursued in large-scale microarray experiments (Dudoit et al. 2003), brain imaging and fMRI analysis (Nichols and Hayasaka 2003), and numerous other anomaly detection applications.

A common theme in such applications is that the errors are *correlated*, and that the signal vectors are believed to be *sparse*: the number of non-zero (or large) components in μ is small compared to the number of tests performed. In the cybersecurity context, while a very large number of data streams are monitored, typically only just a few of them will be experiencing problems at any time, barring large-scale outages or distributed denial-of-service attacks. Under such sparsity assumptions, it is natural to ask if and when one can reliably (1) detect the signals and (2) recover the support set S. In this text, we explore both the *detection* and the *support recovery* problems. More precisely, we are interested in the theoretical feasibility of both problems, and seek minimal conditions under which these problems can be consistently solved in large dimensions.

Model (1.1) is simple yet ubiquitous. Consider the linear model

$$Y = X\mu + \xi,$$

where μ is a p-dimensional vector of regression coefficients of interest to be inferred from observations of X and Y. If the design matrix X is of full column rank,[1] then the ordinary least squares (OLS) estimator of μ can be formed

$$\widehat{\mu} = \left(X'X\right)^{-1} X'Y = \mu + \epsilon, \tag{1.2}$$

where $\epsilon := (X'X)^{-1} X'\xi$. Hence, we recover the generic problem (1.1). Signal detection is therefore equivalent to the problem of testing the global null model, and support recovery problem corresponds to the fundamental problem of variable selection.

Note that the components of the observation vector x (and equivalently, the noise ϵ) in (1.1) need not be independent. In the linear regression example, even when the components of the noise term ξ are independent, those of the OLS estimator (1.2) need not be, except in the case of orthogonal designs. Indeed, in practice, independence is the exception rather than the rule. Therefore, a general theory of feasibility must address the role of the *error-dependence* structure in such testing and support estimation problems. It is also important to identify practical and/or optimal procedures that attain the performance limits in independent as well as dependent cases, as soon as the problems become theoretically feasible. We address both themes in this text.

1.2 Genome-Wide Association Studies and the Chi-Square Model

The second data model we analyze is the high-dimensional chi-square model,

$$x(i) \sim \chi_\nu^2 (\lambda(i)), \quad i = 1, \ldots, p, \tag{1.3}$$

where the data $x(i)$'s follow independent (non-central) chi-square distributions with ν degrees of freedom and non-centrality parameter $\lambda(i)$.

---

[1] This, of course, requires that we have more samples than dimensions, i.e., $n > p$. Nevertheless, multiplicity of tests is still present when p itself is large—the multiple testing problem is by no means exclusive to situations where $p \gg n$.

Table 1.1 Tabulated counts of genotype-phenotype combinations in a genetic association test

# Observations	Genotype		Total by phenotype
	Variant 1	Variant 2	
Cases	O_{11}	O_{12}	n_1
Controls	O_{21}	O_{22}	n_2

Model (1.3) is motivated by large-scale categorical variable screening problems, typified by GWAS where millions of genetic factors are examined for their potential influence on phenotypic traits.

In a GWAS with a case-control design, a total of n subjects are recruited, consisting of n_1 subjects possessing some defined traits, and n_2 subjects without the traits serving as controls. The genetic compositions of the subjects are then examined for variations known as SNP at an array of p genomic marker locations, and compared between the case and the control group. These physical traits are commonly referred to as *phenotypes*, and the genetic variations are known as *genotypes* (Table 1.1).

Focusing on one specific genomic location, the counts of observed genotypes, if two variants are present, can be tabulated as follows. Researchers test for associations between the genotypes and phenotypes using, for example, the Pearson chi-square test with statistic

$$x = \sum_{j=1}^{2} \sum_{k=1}^{2} \frac{(O_{jk} - E_{jk})^2}{E_{jk}}, \tag{1.4}$$

where $E_{jk} = (O_{j1} + O_{j2})(O_{1k} + O_{2k})/n$.

Under the mild assumption that the counts O_{jk}'s follow a multinomial distribution (or a product-binomial distribution, if we decide to condition on one of the marginals), the statistic x in (1.4) can be shown to have an approximate $\chi^2(\lambda)$ distribution with $\nu = 1$ degree of freedom at large sample sizes (see, e.g., classical results in Ferguson 2017; Agresti 2018). Independence between the genotypes and phenotypes would imply a non-centrality parameter λ value of zero; if dependence exists, we would have a non-zero λ where its value depends on the underlying multinomial probabilities. More generally, if we have a J phenotypes and K genetic variants, assuming a $J \times K$ multinomial distribution, the statistic will follow approximately a $\chi_\nu^2(\lambda)$ distribution with $\nu = (J - 1)(K - 1)$ degrees of freedom, when sample sizes are large.

The same asymptotic distributional approximations also apply to the likelihood ratio statistic, and many other statistics under slightly different modeling assumptions (Gao et al. 2019). These association tests are performed at each of the p SNP marker locations throughout the whole genome, and we arrive at p statistics having approximately (non-central) chi-square distributions, $\chi_{\nu(i)}^2 (\lambda(i))$, for $i = 1, \ldots, p$, where $\lambda = (\lambda(i))_{i=1}^{p}$ is the p-dimensional non-centrality parameter.

Although the number of tested genomic locations p can sometimes exceed 10^5 or even 10^6, it is often believed that only a small set of genetic locations have tangible influences on the outcome of the disease or the trait of interest. Under the stylized assumption of sparsity, λ is assumed to have s non-zero components, with s being much smaller than the problem dimension p. The goal of researchers is again twofold: (1) to test if $\lambda(i) = 0$ for all i, and (2) to estimate the set $S = \{i : \lambda(i) \neq 0\}$. In other words, we look to first determine if there are *any* genetic variations associated with the disease, and if there are associations, we want to locate them.

The chi-square model (1.3) also plays an important role in analyzing variable screening problems under omnidirectional alternatives. A primary example is multiple testing under two-sided alternatives in the additive error model (1.1) where the errors ϵ are assumed to have standard normal distributions.

Under two-sided alternatives, unbiased test procedures call for rejecting the hypothesis $\mu(i) = 0$ at locations where observations have large absolute values or, equivalently, large squared values. Taking squares on both sides of (1.1), we arrive at Model (1.3) with non-centrality parameters $\lambda(i) = \mu^2(i)$ and degree-of-freedom parameter $\nu = 1$. In this case, the support recovery problem is equivalent to locating the set of observations with mean shifts, $S = \{i : \mu(i) \neq 0\}$, where the mean shifts could take place in both directions.

Therefore, a theory for the chi-square model (1.3) naturally lends itself to the study of two-sided alternatives in the Gaussian additive error model (1.1). In comparing such results with existing theory on one-sided alternatives, we will be able to quantify if, and how much of a price has to be paid for the additional uncertainty when we have no prior knowledge on the direction of the signals.

1.3   Contents

Important notions and definitions in high-dimensional testing problems are recalled in Chap. 2. We review related literature as well as key concepts and technical results used in our subsequent analyses.

In Chap. 3, we study the sparse signal detection and support recovery problems for the additive error model (1.1) when components of the noise term ϵ are independent standard Gaussian random variables. In particular, we point out several new *phase transitions* in signal detection problems, and provide a unified account of recently discovered phase transitions in support recovery problems. These results show that as the dimension $p \to \infty$, the tasks of detecting the existence of signals or identifying the support set S are either doable or impossible depending on the sparsity and signal sizes of the problems. We also identify commonly used procedures that attain the performance limits in both detection and support recovery problems.

Both the Gaussianity assumption and the independence assumption are relaxed in Chap. 4. Established are the necessary and sufficient conditions for exact support recovery in the high-dimensional asymptotic regime for the large class of thresholding procedures. This is a major theoretical contribution of our approach, which solves and expands on open problems in the recent literature (see Butucea et al. 2018; Gao and Stoev 2020). The analysis of support recovery problem is intimately related to a *concentration of maxima* phenomena in the analysis of extremes. The latter concept is key to understanding the role played by dependence in the phase-transition phenomena of high-dimensional testing problems. In Chap. 5, we study the universality of the phase-transition phenomenon in exact support recovery. We do so by first establishing the finite-sample Bayes optimality and sub-optimality of thresholding procedures. This, combined with the results from Chap. 4, culminates in asymptotic minimax characterizations of the phase-transition phenomenon in exact support recovery across all procedures for a large class of dependence structures.

The dependence condition defined by the concentration of maxima concepts is further demystified in Chap. 6 for Gaussian errors. We offer a complete characterization of the concentration of maxima phenomenon, known as uniform relative stability, in terms of the covariance structures of the Gaussian arrays. This result may be of independent interest since it relates to the so-called *superconcentration* phenomenon coined by Chatterjee (2014). See also, Gao and Stoev (2020), Kartsioukas et al. (2019).

Chapter 7 returns to high-dimensional multiple testing problems and studies the chi-square model (1.3) inspired by the marginal association screening problems. We establish four new phase-transition-type results in the chi-square model, and illustrate their practical implications in the GWAS application. Our theory enables us to explain the long-standing empirical observation that small perturbations in the frequency and penetrance of genetic variations lead to drastic changes in the discoverability in genetic association studies.

Chapter 2
Risks, Procedures, and Error Models

We establish the background necessary for the study of sparse signal detection and support recovery problems in this chapter. Sections 2.1 and 2.2 provide a refresher on the definitions of statistical risks and some commonly used statistical procedures. Section 2.3 describes the asymptotic regime under which we analyze these procedures, and reviews the related literature in high-dimensional statistics. We discuss in Sect. 2.4 the connections among the risk metrics, and point out some common fallacies. The remaining sections collect the technical preparations for this text. Section 2.5 defines an important class of error distributions which will be analyzed in detail in later chapters. Section 2.6 introduces the concepts of concentration of maxima, which plays a crucial role in the analysis of high-dimensional support recovery problems. Finally, in Sect. 2.7, we gather well-known but indispensable facts about Gaussian distributions.

2.1 Statistical Risks

We define the statistical risk metrics for signal detection and signal support recovery problems in this section. Formally, we denote a statistical procedure, i.e., measurable function of the data, as $\mathcal{R} = \mathcal{R}(x)$. In the testing context, a procedure $\mathcal{R}$ produces a binary decision T that represents our judgment on the presence or absence of a signal. In the support recovery problem, a procedure $\mathcal{R}$ produces an index set $\widehat{S}$ that represents our estimate of the signal support. The statistical risks are then suitable functionals of T and $\widehat{S}$ in respective contexts.

© The Author(s), under exclusive license to Springer Nature Switzerland AG 2021
Z. Gao and S. Stoev, *Concentration of Maxima and Fundamental Limits in High-Dimensional Testing and Inference*,
SpringerBriefs in Probability and Mathematical Statistics,
https://doi.org/10.1007/978-3-030-80964-5_2

Signal detection. Recall that in sparse signal detection problems, our goal is to come up with a procedure, $\mathcal{R}(x)$, such that the null hypothesis is rejected if the data x is deemed incompatible with the null. In the additive error models context (1.1), we wish to tell apart two hypotheses

$$\mathcal{H}_0 : \mu(i) = 0, \ i = 1, \ldots, p, \quad \text{v.s.} \quad \mathcal{H}_1 : \mu(i) \neq 0, \ \text{for some } i \in \{1, \ldots, p\}, \tag{2.1}$$

based on the p-dimensional observation x. Similarly, in the chi-square model (1.3), we look to test

$$\mathcal{H}_0 : \lambda(i) = 0, \ i = 1, \ldots, p, \quad \text{v.s.} \quad \mathcal{H}_1 : \lambda(i) \neq 0, \ \text{for some } i \in \{1, \ldots, p\}. \tag{2.2}$$

Since the decision is binary, we may write the outcome of the procedure in the form of an indicator function, $T(\mathcal{R}(x)) \in \{0, 1\}$, where $T = 1$ if the null is to be rejected in favor of the alternative, and 0 if we fail to reject the null. The Type I and Type II errors of the procedure, i.e., the probability of wrong decisions under the null hypothesis $\mathcal{H}_0$ and alternative hypothesis $\mathcal{H}_1$, respectively, are defined as

$$\alpha(\mathcal{R}) := \mathbb{P}_{\mathcal{H}_0}\left(T(\mathcal{R}(x)) = 1\right) \quad \text{and} \quad \beta(\mathcal{R}) := \mathbb{P}_{\mathcal{H}_1}\left(T(\mathcal{R}(x)) = 0\right). \tag{2.3}$$

The Neyman–Pearson framework of hypothesis testing then seeks tests that minimize the Type II error of the test, while controlling the Type I error of the test at low levels. We are particularly interested in the sum of the two errors,

$$\text{risk}^{\mathrm{D}}(\mathcal{R}) := \alpha(\mathcal{R}) + \beta(\mathcal{R}), \tag{2.4}$$

which shall be referred to as the risk of signal detection (of the procedure $\mathcal{R}$). It is trivial that a small risk^{D} would imply both small Type I and Type II errors of the procedure.

Signal support recovery. Turning to support recovery problems, our goal is to design a procedure that produces a set estimate $\widehat{S}(\mathcal{R}(x))$ of the true index set of relevant variables S. For example, in the sparse additive error model (1.1), we aim to estimate $S = \{i : \mu(i) \neq 0\}$, while in the sparse chi-square model (1.3) the goal is to estimate $S = \{i : \lambda(i) \neq 0\}$. For simplicity of notation, we shall write $\widehat{S}$ for the support estimator $\widehat{S}(\mathcal{R}(x))$.

For a given procedure $\mathcal{R}$, its false discovery rate (FDR) and false non-discovery rate (FNR) are defined, respectively, as

$$\text{FDR}(\mathcal{R}) := \mathbb{E}\left[\frac{|\widehat{S} \setminus S|}{\max\{|\widehat{S}|, 1\}}\right] \quad \text{and} \quad \text{FNR}(\mathcal{R}) := \mathbb{E}\left[\frac{|S \setminus \widehat{S}|}{\max\{|S|, 1\}}\right], \tag{2.5}$$

where the maxima in the denominators resolve the possible division-by-0 problem. Roughly speaking, FDR measures the expected fraction of false findings, while FNR describes the proportion of Type II errors among the true signals, and reflects the average marginal power of the procedure.

A more stringent criterion for false discovery is the family-wise error rate (FWER), defined to be the probability of reporting at least one finding not contained in the true index set. Correspondingly, a more stringent criterion for false non-discovery is the family-wise non-discovery rate (FWNR), i.e., the probability of missing at least one signal in the true index set. That is,

$$\mathrm{FWER}(\mathcal{R}) := 1 - \mathbb{P}[\widehat{S} \subseteq S] \quad \text{and} \quad \mathrm{FWNR}(\mathcal{R}) := 1 - \mathbb{P}[S \subseteq \widehat{S}]. \qquad (2.6)$$

We introduce five different statistical risk metrics, each having different asymptotic limits in the support recovery problems as we will see in Chap. 3. Following Arias-Castro and Chen (2017), we define the risk for *approximate* support recovery as

$$\mathrm{risk}^{\mathrm{A}}(\mathcal{R}) := \mathrm{FDR}(\mathcal{R}) + \mathrm{FNR}(\mathcal{R}). \qquad (2.7)$$

Analogously, we define the risk for *exact* support recovery as

$$\mathrm{risk}^{\mathrm{E}}(\mathcal{R}) := \mathrm{FWER}(\mathcal{R}) + \mathrm{FWNR}(\mathcal{R}). \qquad (2.8)$$

Two closely related measures of success in the exact support recovery risk are the probability of exact recovery,

$$\mathbb{P}[\widehat{S} = S] = 1 - \mathbb{P}[\widehat{S} \neq S], \qquad (2.9)$$

and the Hamming loss

$$H(\widehat{S}, S) := |\widehat{S} \triangle S| = \sum_{i=1}^{p} \left| \mathbb{1}_{\widehat{S}}(i) - \mathbb{1}_{S}(i) \right|, \qquad (2.10)$$

which counts the number of mismatches between the estimated and true support sets.

The relationship between probability of support recovery $\mathbb{P}[\widehat{S} = S]$, exact support recovery risk $\mathrm{risk}^{\mathrm{E}}$, and the expected Hamming loss $\mathbb{E}[H(\widehat{S}, S)]$ will be discussed in Sect. 2.4.

Notice that all risk metrics introduced so far penalize false discoveries and missed signals somewhat symmetrically—the approximate support recovery risk combines proportions of errors, the exact support recovery risk combines probabilities of errors, and the Hamming loss increments the risk by one regardless of the types of errors made. In applications, however, attitudes toward false discoveries and missed signals are often asymmetric. In the example of GWAS, where the number of candidate locations p could be in the millions, and a class imbalance between the number of nulls and signals exists, researchers are typically interested in the marginal (location-wise) power of discovery, while exercising stringent (family-wise) false discovery control. These types of asymmetric considerations, while important in applications, have not been studied theoretically. For example, the GWAS application motivates the *exact–approximate* support recovery risk, which weighs both the family-wise error rate and the marginal power of discovery:

$$\text{risk}^{\text{EA}}(\mathcal{R}) := \text{FWER}(\mathcal{R}) + \text{FNR}(\mathcal{R}). \tag{2.11}$$

The somewhat cumbersome name and notation are chosen to reflect the asymmetry in dealing with the two types of errors in support recovery. Namely, when the risk metric (2.11) vanishes, we have "exact false discovery control, and approximate false non-discovery control" asymptotically.

Analogously, we consider the *approximate–exact* support recovery risk

$$\text{risk}^{\text{AE}}(\mathcal{R}) := \text{FDR}(\mathcal{R}) + \text{FWNR}(\mathcal{R}), \tag{2.12}$$

which places more emphasis on non-discovery control over false discovery.

Theoretical limits and performance of procedures in support recovery problems will be studied in terms of the five risk metrics (2.7), (2.8), (2.9), (2.11), and (2.12), in Chaps. 3, 4, and 7. We are particularly interested in fundamental limits of signal detection and support recovery problems in terms of these metrics, as well as the optimality of commonly used procedures in high-dimensional settings.

2.2 Statistical Procedures

We review some popular procedures for signal detection and signal support recovery tasks in this section.

Signal detection. One of the commonly used statistics in sparse signal detection problems such as (2.1) and (2.2) are the L_q norms of the observations x,

$$L_q(x) = \left(\sum_{i=1}^{p} |x(i)|^q \right)^{1/q}. \tag{2.13}$$

Typical choices of q include $q = 1, 2$ and ∞, where $L_\infty(x)$ is interpreted as the limit of $L_q(x)$ norms as $q \to \infty$, and is equivalent to $\max_i |x(i)|$. Test procedures based on (2.13) may then be written as $T(\mathcal{R}(x)) = \mathbb{1}_{(t,+\infty)}(L_q(x))$, where the cutoff t can be chosen to control the Type I error at desired levels.

While (2.13) measures the deviation of the data from the origin in an omnidirectional manner, statistics that are tailored to the alternatives can be used in the hopes of power improvement if the directions of the alternatives are known. For example, in the additive error model (1.1), suppose we want to test for positive mean shifts, i.e., one-sided alternative

$$\mathcal{H}_1 : \mu(i) > 0, \text{ for some } i \in \{1, \dots, p\}. \tag{2.14}$$

Then, one might consider monitoring the sum (or equivalently, the arithmetic average) of the observations,

$$T(x) := \sum_{i=1}^{p} x(i), \tag{2.15}$$

or the maximum of the observations,

$$M(x) := \max_{i=1,\dots,p} x(i). \tag{2.16}$$

Other tests based on the empirical CDF are also available. Assuming the same one-sided alternative, let

$$q(i) = 1 - \sup\{F_i(y) : y < x(i)\}, \quad i = 1, \dots, p \tag{2.17}$$

be the p-values of the individual observations, where F_i is the CDF of the i-th component $x(i)$ under $\mathcal{H}_0$. We define empirical CDF of the p-values as

$$\widehat{F}_p(t) = \frac{1}{p} \sum_{i=1}^{p} \mathbb{1}_{[0,t]}(q(i)). \tag{2.18}$$

Viewed as random elements in the space of càdlàg functions with the Skorohod J_1 topology, the centered and scaled CDFs converge weakly to a Brownian bridge,

$$\left\{\sqrt{p}\left(\widehat{F}_p(t) - t\right)\right\}_{t \in [0,1]} \implies \{\mathbb{B}(t)\}_{t \in [0,1]}, \quad \text{as } p \to \infty,$$

under the global null $\mathcal{H}_0$ and mild continuity assumptions on the F_i's (Skorokhod 1956). Therefore, goodness-of-fit statistics such as Kolmogorov–Smirnov distance (Smirnov 1948), Cramer–von Mises-type statistics (Cramér 1928; Anderson and Darling 1952) that measure the departure from this limiting behavior can be used for testing $\mathcal{H}_0$ against $\mathcal{H}_1$. Of particular interest is the higher criticism (HC) statistic, first proposed by Tukey (1976),

$$HC(x) = \max_{0 \le t \le \alpha_0} \frac{\widehat{F}_p(t) - t}{\sqrt{t(1-t)/p}}. \tag{2.19}$$

Each of the above statistics L_q, S, M, or HC gives rise to a decision rule, whereby the null hypothesis is rejected if the statistic exceeds a suitably calibrated threshold. The choice of the threshold is typically determined based on large-sample limit theorems. For example, as shown in Theorem 1.1 of Donoho and Jin (2004), under the null hypothesis

$$\frac{HC(x)}{\sqrt{2 \log \log(p)}} \longrightarrow 1, \quad \text{in probability,}$$

as $p \to \infty$. Thus, one decision rule is to reject $\mathcal{H}_0$, if $HC(x) > t(p, \alpha_p)$, where $t(p, \alpha_p) = \sqrt{2 \log \log(p)}(1 + o(1))$. As we will see, this yields an optimal signal detection procedure (see also Theorem 1.2 in Donoho and Jin 2004). The performance of these statistics in high-dimensional sparse signal detection problems will be reviewed in Sect. 2.3, and analyzed in Chap. 3.

Signal support recovery. In signal support recovery tasks, we shall study the performance of five procedures, all of which belong to the broad class of thresholding procedures.

Definition 2.1 (*Thresholding procedures*) A thresholding procedure for estimating the support $S := \{i \, : \, \lambda(i) \neq 0\}$ is one that takes on the form

$$\widehat{S} = \{i \mid x(i) \geq t(x)\}, \tag{2.20}$$

where the threshold $t(x)$ may depend on the data x.

Examples of thresholding procedures include ones that aim to control FWER (2.6)—Bonferroni's (Dunn 1961), Sidák's (Šidák 1967), Holm's (Holm 1979), and Hochberg's procedure (Hochberg 1988)—as well as procedures that target FDR (2.5), such as the Benjamini–Hochberg Benjamini and Hochberg (1995) and the Barber–Candès procedure (Barber and Candès 2015). Indeed, the class of thresholding procedures (2.20) is so general that it contains most (but not all) of the statistical procedures in the multiple testing literature.

Under the assumption that the data $x(i)$'s under the null have a common marginal distribution F, we review five thresholding procedures for support recovery, starting with the well-known Bonferroni's procedure which aims at controlling family-wise error rates.

Definition 2.2 (*Bonferroni's procedure*) Bonferroni's procedure with level α is the thresholding procedure that uses the threshold

$$t_p = F^{\leftarrow}(1 - \alpha/p), \tag{2.21}$$

where $F^{\leftarrow}(u) = \inf \{x \, : \, F(x) \geq u\}$ is the generalized inverse function.

The Bonferroni procedure is deterministic, i.e., non-data-dependent, and only depends on the dimension of the problem and the null distribution. A closely related procedure is Sidák's procedure (Šidák 1967), which is a more aggressive (and also deterministic) thresholding procedure that uses the threshold

$$t_p = F^{\leftarrow}((1 - \alpha)^{1/p}). \tag{2.22}$$

The third procedure, strictly more powerful than Bonferroni's, is the so-called Holm's procedure (Holm 1979). On observing the data x, its coordinates can be ordered from largest to smallest $x(i_1) \geq x(i_2) \geq \ldots \geq x(i_p)$, where $(i_1, \ldots, i_p)$ is a permutation of $\{1, \ldots, p\}$. Denote these order statistics as $x_{[1]}, x_{[2]}, \ldots, x_{[p]}$.

Definition 2.3 (*Holm's procedure*) Let i^* be the largest index such that

$$\overline{F}(x_{[i]}) \leq \alpha/(p - i + 1), \quad \text{for all } i \leq i^*.$$

Holm's procedure with level α is the thresholding procedure with threshold

$$t_p(x) = x_{[i^*]}. \tag{2.23}$$

In contrast to the Bonferroni procedure, Holm's procedure is data-dependent. A closely related, more aggressive (and also data-dependent) thresholding procedure is Hochberg's procedure (Hochberg 1988). It replaces the index i^* in Holm's procedure with the largest index such that

$$\overline{F}(x_{[i]}) \leq \alpha/(p - i + 1).$$

Notice that both Holm's and Hochberg's procedures compare p-values to the same thresholds $\alpha/(p - i + 1)$. However, Holm's procedure only rejects the set of hypotheses whose p-values are all smaller than their respective thresholds. On the other hand, Hochberg's procedure rejects the set of hypotheses as long as the largest of their p-values fall below its threshold, and therefore can be more powerful than Holm's procedure.

It can be shown that both Bonferroni's and Holm's procedures control FWER at their nominal levels, regardless of dependence in the data (Holm 1979). In contrast, Sidák's and Hochberg's procedures control FWER at nominal levels when data are independent (Šidák 1967; Hochberg 1988).

Last but not least, we review the BH procedure, which aims at controlling FDR in (2.5), proposed by Benjamini and Hochberg (1995).

Recall the order statistics of our observations are: $x_{[1]} \geq x_{[2]} \geq \ldots \geq x_{[p]}$.

Definition 2.4 (*Benjamini–Hochberg's procedure*) Let i^* be the largest index such that

$$\overline{F}(x_{[i]}) \leq \alpha i/p.$$

The Benjamini–Hochberg (BH) procedure with level α is the thresholding procedure with threshold

$$t_p(x) = x_{[i^*]}. \tag{2.24}$$

The BH procedure is shown to control the FDR at level α when the $x(i)$'s are independent (Benjamini and Hochberg 1995). Variations of this procedure have been proposed to control the FDR under certain models of dependent observations (Benjamini and Yekutieli 2001).

The performance of these procedures in high-dimensional sparse signal support recovery problems will be reviewed in Sect. 2.3, and analyzed in Chaps. 3, 4, and 7.

2.3 Related Literature and Our Contributions

We look to derive useful asymptotic approximations for high-dimensional problems, and analyze the aforementioned procedures in the regime where the dimensionality of the observations diverges. Throughout this text, we consider triangular arrays of observations as described in Models (1.1) and (1.3), and study the performance of various procedures in the signal detection and support recovery tasks when

$$p \to \infty.$$

The criteria for success and failure in support recovery problems under this high-dimensional asymptotic regime are defined as follows.

Definition 2.5 We say a sequence of procedures $\mathcal{R} = \mathcal{R}_p$ succeeds asymptotically in the detection problem (and, respectively, exact, exact–approximate, approximate–exact, and approximate support recovery problem) if

$$\mathrm{risk}^{\mathrm{P}}(\mathcal{R}) \to 0, \quad \text{as} \quad p \to \infty, \tag{2.25}$$

where P = D (respectively, E, EA, AE, A).

Conversely, we say the exact support recovery fails asymptotically in the detection problem (and, respectively, exact, exact–approximate, approximate–exact, and approximate support recovery problem) if

$$\liminf \mathrm{risk}^{\mathrm{P}}(\mathcal{R}) \geq 1, \quad \text{as} \quad p \to \infty, \tag{2.26}$$

where P = D (respectively, E, EA, AE, A).

The choice of the constant 1 in Definition (2.26) allows us to declare failure for trivial testing procedures. For example, trivial deterministic procedures that always reject and ones that always fail to reject both have statistical risks 1 in either the detection or the support recovery problem. Similarly, a trivial randomized procedure that rejects the nulls uniformly at random also has risk of 1, and is declared as a failure in both problems.

Signal detection. The asymptotic behavior of the statistical risk in signal detection problems (2.4) in high dimensions was first studied by Yuri Izmailovich Ingster in the context of sparse additive models (1.1) with independent and Gaussian components. Specifically, Ingster (1998) considered the behavior of the theoretically optimal likelihood ratio (LR) test in the high-dimensional asymptotic regime, where the dimension p grows to infinity. Then, under certain parameterization of the size and sparsity of the signal μ, a dichotomy exists: either $\mathrm{risk}^D(\mathcal{R})$ vanishes as $p \to \infty$ where $\mathcal{R}$ is the LR test, or $\liminf_{p\to\infty} \mathrm{risk}^D(\mathcal{R}) = 1$ for any procedure. The precise signal size and sparsity parameterizations as well as the so-called *detection boundary* discovered by Ingster are described in Chap. 3.

The LR test, unfortunately, relies on the knowledge of the signal sparsity and signal sizes which are not available in practice. The sparsity and signal size agnostic statistic HC in (2.19) was identified to attain such optimal performance limits in sparse Gaussian models in Donoho and Jin (2004). A modified goodness-of-fit test statistic in Zhang (2002) and two statistics based on thresholded-L_1 and L_2 norms in Zhong et al. (2013) were also shown to be asymptotically optimal in the detection problem. Recent studies have also focused on the behavior of detection risk (2.4) in dense and scale mixture models (Cai et al. 2011), under general distributional assumptions (Cai and Wu 2014; Arias-Castro and Wang 2017), as well as when the errors are dependent (Hall and Jin 2010). A comprehensive review focusing on the role of HC in detection problems can be found in Donoho and Jin (2015). The very recent contribution of Li and Fithian (2020) shows exciting new developments on the detection problem in a more realistic regime than the ones previously studied in the literature. It shows that the max-statistic begins to attain the optimal boundary and is on par with HC (cf Table 1, therein). Notwithstanding the extensive literature on the detection problem, the performances of simple statistics such as L_q norms (2.13) and sums (2.15), to the best of our knowledge, have only been sparingly documented. We gather relevant results in Chap. 3, and make several new contributions on the performance of several statistics commonly used in practice.

Exact support recovery. There is a wealth of literature on the so-called sparsistency (i.e., $\mathbb{P}[\widehat{S} = S] \to 1$ as $p \to \infty$) problem in the regression context. Sparsistency problems were pursued, among many others, by Zhao and Yu (2006), Wasserman and Roeder (2009) in the high-dimensional regression setting (where the number of samples $n \ll p$), and by Meinshausen and Bühlmann (2006) in graphical models. Although there have been numerous studies on the sufficient conditions for sparsistency, efforts on necessary conditions have been scarce. Notable exceptions include Wainwright (2009a, b), Comminges and Dalalyan (2012) in regression problems. We refer the reader to the recent book by Wainwright (2019) (and, in particular, the bibliographical sections of Chaps. 7 and 15 therein) for a comprehensive review.

Elaborate asymptotic minimax-optimality results under the Hamming loss were derived for methods proposed in Ji and Jin (2012), Jin et al. (2014) for regression problems. More recently, Butucea et al. (2018) also obtained similar minimax-optimality results for a specific procedure in the Gaussian additive error model (1.1) in terms of the expected Hamming loss.

Nevertheless, two important questions remained unanswered. Namely, precise phase-transition-type results for the exact support recovery risk (2.8) and for the support recovery probability (2.9) have not been established. And secondly, performance of commonly used statistical procedures reviewed in Sect. 2.2 in terms of these risk metrics have not been studied. Some of our main contributions in this text are the solutions to these problems, presented in Chaps. 3 and 4. Specifically, we show that the Bonferroni thresholding procedure (among others) is asymptotically optimal for the exact support recovery problem in (1.1) under broad classes of error distributions. Furthermore, a phase transition in the exact support recovery problem for thresholding procedures is established under broad dependence conditions on the errors using the concentration of maxima phenomenon (Chap. 4). We also establish finite-sample Bayes optimality and sub-optimality results for these procedures under independence, and by extension arrive at minimax-optimality results for the exact support recovery problem (Chap. 5).

The landscape of the fundamental statistical limits in support estimation is yet to be fully charted. We conjecture, however, that the general concentration of maxima phenomenon will lead to its complete solution under very broad error-dependence scenarios.

Approximate support recovery. The performance limits of FDR-controlling procedures in the support recovery problem have been actively studied in recent years. The asymptotic optimality of the Benjamini–Hochberg procedure was analyzed under decision theoretic frameworks in Genovese and Wasserman (2002); Bogdan et al. (2011); Neuvial and Roquain (2012), with main focus on location/scale models. In particular, these papers show that the statistical risks of the procedures come close to those of the oracle procedures under suitable asymptotic regimes. Strategies for dealing with multiple testing under general distributional assumptions can be found in, e.g., Efron (2004), Storey (2007), Sun and Cai (2007). The two-sided alternative in the additive error model was featured as the primary example in Sun and Cai (2007).

In the additive error model (1.1) under independent Gaussian errors and one-sided alternatives (2.14), Arias-Castro and Chen (2017) showed that a phase transition exists for the approximate support recovery risk (2.7). The BH procedure (Benjamini and Hochberg 1995) and the Barber–Candès procedure (Barber and Candès 2015) were identified to be asymptotically optimal. However, Arias-Castro and Chen (2017), as all related work so far, assumed the non-nulls to follow a common alternative distribution. We derive a new phase-transition result that relaxes this assumption on the alternatives in Chap. 3.

Asymmetric statistical risks. Although weighted sums of false discovery and non-discovery have been studied in the literature mentioned above, the case of simultaneous family-wise error control and marginal, location-wise power requirements has not been previously considered. As a result, asymmetric statistical risks such as (2.11) and (2.12) have not been investigated. As argued in Sect. 2.1, the properties of these asymmetric risks are of important practical concern in applications such as GWAS. We study the asymptotic behavior of these risks in Chaps. 3 and 7 of this text.

Chi-square models and GWAS. The high-dimensional chi-square model (1.3) seemed to have received little attention in the literature. While the sparse signal detection problem in the chi-square model has been studied (Donoho and Jin 2004), to the best of our knowledge, asymptotic limits of the support recovery problems have not been studied. The chi-squared model and the motivating GWAS application are analyzed in Chap. 7. The results obtained therein help us explain a phenomenon in GWAS where statistical power decays sharply as function of sample size when the latter is in a small region known as the *steep part of the power curve*. This empirical fact has long been observed by statistical geneticists but has not been mathematically quantified. Gao et al. (2019) provide further details on the power and design in GWAS as well as an accompanying interactive statistical software (Gao 2019).

2.4 Relationships Between the Asymptotic Risks

We now elaborate on the relationship between statistical risks, as promised in Sect. 2.1. The first lemma concerns the asymptotic relationship between the probability of exact recovery (2.9) and the risk of exact support recovery (2.8).

Lemma 2.1 *For any sequence of procedures for support recovery* $\mathcal{R} = \mathcal{R}_p$, *we have*

$$\mathbb{P}[\widehat{S} = S] \to 1 \iff risk^E(\mathcal{R}) \to 0, \tag{2.27}$$

and

$$\mathbb{P}[\widehat{S} = S] \to 0 \implies \liminf risk^E(\mathcal{R}) \geq 1, \tag{2.28}$$

as $p \to \infty$. *Dependence on* p *and* $\mathcal{R}$ *was suppressed for notational convenience.*

Proof (*Lemma* 2.1) Notice that $\{\widehat{S} = S\}$ implies $\{\widehat{S} \subseteq S\} \cap \{\widehat{S} \supseteq S\}$, and therefore we have for every fixed p,

$$risk^E = 2 - \mathbb{P}[\widehat{S} \subseteq S] - \mathbb{P}[S \subseteq \widehat{S}] \leq 2 - 2\mathbb{P}[\widehat{S} = S]. \tag{2.29}$$

On the other hand, since $\{\widehat{S} \neq S\}$ implies $\{\widehat{S} \not\subseteq S\} \cup \{\widehat{S} \not\supseteq S\}$, we have for every fixed p,

$$1 - \mathbb{P}[\widehat{S} = S] = \mathbb{P}[\widehat{S} \neq S] \leq 2 - \mathbb{P}[\widehat{S} \subseteq S] - \mathbb{P}[S \subseteq \widehat{S}] = risk^E. \tag{2.30}$$

Relation (2.27) follows from (2.29) and (2.30), and Relation (2.28) from (2.30). □

By virtue of Lemma 2.1, it is sufficient to study the probability of exact support recovery $\mathbb{P}[\widehat{S} = S]$ in place of $risk^E$, if we are interested in the asymptotic properties of the risk in the sense of (2.25) and (2.26).

Keen readers must have noticed the asymmetry in Relation (2.28) when we discussed the relationship between the exact support recovery risk (2.8) and the probability of exact support recovery (2.9). While a trivial procedure that never rejects and a procedure that always rejects both have $risk^E$ equal to 1, the converse is not true. For example, it is possible that a procedure selects the true index set S with probability $1/2$, but otherwise makes one false inclusion *and* one false omission simultaneously. In this case, the procedure will have

$$risk^E = 1, \quad \text{and} \quad \mathbb{P}[\widehat{S} = S] = 1/2,$$

showing that the converse of Relation (2.28) is in fact false.

The same argument applies to risk^A: a procedure may select the true index set S with probability $1/2$, but makes enough false inclusions and omissions the rest of the time, so that risk^A is equal to one. Therefore, although the class of methods with risks equal to or exceeding 1 certainly contains the trivial procedures that we mentioned, they are not necessarily "useless" as some researchers have claimed (cf. Remark 2 in Arias-Castro and Chen 2017).

Upper and lower bounds for FDR and FNR can be immediately derived by replacing the numerators in (2.5) with the Hamming loss,

$$\mathbb{E}\left[\frac{H(\widehat{S}, S)}{\max\{|\widehat{S}|, |S|, 1\}}\right] \leq \text{FDR} + \text{FNR} \leq \mathbb{E}\left[\frac{H(\widehat{S}, S)}{\max\{\min\{|\widehat{S}|, |S|\}, 1\}}\right]. \qquad (2.31)$$

Therefore, it is sufficient, but not necessary, that the Hamming loss vanishes in order to have vanishing approximate support recovery risks (2.7).

Turning to the relationship between the probability of exact support recovery (2.9) and Hamming loss (2.10), we point out a natural lower bound of the former using the expectation of the latter,

$$\mathbb{P}[\widehat{S} = S] \geq 1 - \mathbb{E}[H(\widehat{S}, S)] = 1 - \sum_{i=1}^{p} \mathbb{E}\left|\mathbb{1}_{\widehat{S}}(i) - \mathbb{1}_{S}(i)\right|. \qquad (2.32)$$

A key observation in Relation (2.32) is that the expected Hamming loss decouples into p terms, and the dependence of the estimates $\mathbb{1}_{\widehat{S}}(i)$ among the p locations no longer plays any role in the sum. Therefore, studying support recovery problems via the expected Hamming loss is not very informative especially under severe dependence, as the bound (2.32) may become very loose. Vanishing Hamming loss is again sufficient, but not necessary for $\mathbb{P}[\widehat{S} = S]$ or the exact support recovery risk to go to zero.

2.5 The Asymptotic Generalized Gaussian (AGG) Models

We introduce a fairly general class of distributions known as asymptotic generalized Gaussians AGG. We also state some of their tail properties which play important roles in the analysis of phase transitions of high-dimensional testing problems.

Definition 2.6 A distribution F is called asymptotic generalized Gaussian with parameter $\nu > 0$ (denoted AGG(ν)) if

1. $F(x) \in (0, 1)$ for all $x \in \mathbb{R}$ and
2. $\log \overline{F}(x) \sim -\frac{1}{\nu}x^{\nu}$ and $\log F(-x) \sim -\frac{1}{\nu}(-x)^{\nu}$,

where $\overline{F}(x) = 1 - F(x)$ is the survival function, and $a(x) \sim b(x)$ is taken to mean $\lim_{x \to \infty} a(x)/b(x) = 1$.

The AGG models include, for example, the standard Gaussian distribution ($\nu = 2$) and the Laplace distribution ($\nu = 1$) as special cases. Since the requirement is only placed on the tail behavior, this class encompasses a large variety of light-tailed models. This class is commonly used in the literature on high-dimensional testing (Cai et al. 2007; Arias-Castro and Chen 2017).

Proposition 2.1 *The $(1/p)$-th upper quantile of AGG(ν) is*

$$u_p := F^{\leftarrow}(1 - 1/p) \sim (\nu \log p)^{1/\nu}, \quad as \ p \to \infty, \tag{2.33}$$

where $F^{\leftarrow}(q) = \inf_x\{x : F(x) \geq q\}$, $q \in (0, 1)$ is the generalized inverse of F.

Proof (*Proposition* 2.1) By the definition of AGG, for any $\epsilon > 0$, there is a constant $C = C(\epsilon)$ such that for all $x \geq C$, we have

$$-\frac{1}{\nu}x^\nu(1 + \epsilon) \leq \log \overline{F}(x) \leq -\frac{1}{\nu}x^\nu(1 - \epsilon).$$

Therefore, for all $x < x_l := \left((1 + \epsilon)^{-1}\nu \log p\right)^{1/\nu}$, we have

$$-\log p = -\frac{1}{\nu}x_l^\nu(1 + \epsilon) \leq \log \overline{F}(x_l) \leq \log \overline{F}(x), \tag{2.34}$$

and for all $x > x_u := \left((1 - \epsilon)^{-1}\nu \log p\right)^{1/\nu}$, we have

$$\log \overline{F}(x) \leq \log \overline{F}(x_u) \leq -\frac{1}{\nu}x_u^\nu(1 - \epsilon) = -\log p. \tag{2.35}$$

By the definition of generalized inverse,

$$u_p := F^{\leftarrow}(1 - 1/p) = \inf\{x : \overline{F}(x) \leq 1/p\} = \inf\{x : \log \overline{F}(x) \leq -\log p\}.$$

We know from relations (2.34) and (2.35) that

$$[x_u, +\infty) \subseteq \{x : \log \overline{F}(x) \leq -\log p\} \subseteq [x_l, +\infty),$$

and so $x_l \leq u_p \leq x_u$, and the expression for the quantiles follows. $\square$

2.6 Rapid Variation and Relative Stability

The behavior of the maxima of identically distributed random variables has been studied extensively in the extreme value theory literature (see, e.g., Leadbetter et al. 1983; Resnick 2013; Embrechts et al. 2013; De Haan and Ferreira 2007). The concept of rapid variation plays an important role in the light-tailed case.

Definition 2.7 (*Rapid variation*) The survival function of a distribution, $\overline{F}(x) = 1 - F(x)$, is said to be rapidly varying if

$$\lim_{x \to \infty} \frac{\overline{F}(tx)}{\overline{F}(x)} = \begin{cases} 0, & t > 1 \\ 1, & t = 1 \\ \infty, & 0 < t < 1 \end{cases}. \tag{2.36}$$

When $F(x) < 1$ for all finite x, Gnedenko (1943) showed that the distribution F has rapidly varying tails if and only if the maxima of independent observations from F are *relatively stable* in the following sense.

Definition 2.8 (*Relative stability*) Let $\epsilon_p = \left(\epsilon_p(i)\right)_{i=1}^p$ be a sequence of random variables with common marginal distribution F. Define the sequence $(u_p)_{p=1}^\infty$ to be the $(1 - 1/p)$-th generalized quantile of F, i.e.,

$$u_p = F^{\leftarrow}(1 - 1/p). \tag{2.37}$$

The triangular array $\mathcal{E} = \{\epsilon_p, \, p \in \mathbb{N}\}$ is said to have relatively stable (RS) maxima if

$$\frac{1}{u_p} M_p := \frac{1}{u_p} \max_{i=1,\ldots,p} \epsilon_p(i) \overset{\mathbb{P}}{\to} 1, \quad \text{as } p \to \infty. \tag{2.38}$$

In the case of independent and identically distributed $\epsilon_p(i)$'s, Barndorff-Nielsen (1963), Resnick and Tomkins (1973) obtained necessary and sufficient conditions for the *almost sure stability* of maxima, where the convergence in (2.38) holds almost surely. See also Klass (1984) for further sharp results on almost sure stability, and Naveau (2003) for almost sure stability in stationary sequences. Here, we will only need the weaker notion in (2.38) but extend our inquiry to the case of dependent $\epsilon_p(i)$'s.

While relative stability (and almost sure stability) is well understood in the independent case, the role of dependence has not been fully explored. We start this investigation with a small refinement of Theorem 2 in Gnedenko (1943) valid under *arbitrary dependence*.

Proposition 2.2 (Rapid variation and relative stability) *Assume that the array $\mathcal{E}$ consists of identically distributed and possibly dependent random variables with cumulative distribution function F, where $F(x) < 1$ for all finite $x > 0$.*

1. *If F has rapidly varying right tail in the sense of (2.36), then for all $\delta > 0$,*

$$\mathbb{P}\left[\frac{1}{u_p} M_p \le 1 + \delta\right] \ge 1 - \frac{\overline{F}((1+\delta)u_p)}{\overline{F}(u_p)} \to 1. \tag{2.39}$$

2. *If the array $\mathcal{E}$ has independent entries, then it is relatively stable if and only if F has rapidly varying tail, i.e., (2.36) holds.*

Proof (*Proposition* 2.2) By the union bound and the fact that $p\overline{F}(u_p) \leq 1$, we have

$$\mathbb{P}[M_p > (1+\delta)u_p] \leq p\overline{F}((1+\delta)u_p) \leq \frac{\overline{F}((1+\delta)u_p)}{\overline{F}(u_p)}. \qquad (2.40)$$

In view of (2.36) (rapid variation) and the fact that $u_p \to \infty$, as $p \to \infty$, the right-hand side of (2.40) vanishes as $p \to \infty$, for all $\delta > 0$. This completes the proof of (2.39). Part 2 is a re-statement of a classic result dating back to Gnedenko (1943). $\qquad \square$

Remark 2.1 Part (1) of Proposition 2.2 is equivalent to

$$\mathbb{P}\left[\frac{1}{u_p}M_p > 1+\delta_p\right] \longrightarrow 0, \quad \text{as } p \to \infty, \qquad (2.41)$$

for some positive sequence $\delta_p \to 0$. Notice, on the other hand, that if M_p^* is the maximum of p iid variables with distribution F, the relative stability property entails $M_p^*/u_p \to 1$, in probability, as $p \to \infty$. Since the sequence $1+\delta_p \to 1$, Relation (2.41) means that the rate of growth of the maxima M_n in $\mathcal{E}$ cannot be faster than that of the independent maxima M_p^*. This somewhat surprising fact holds regardless of the dependence structure of $\mathcal{E}$ and is solely a consequence of the rapid variation of F.

We demonstrate next that the Gaussian, Exponential, Laplace, and Gamma distributions all have rapidly varying tails.

Example 2.1 (*Generalized AGG*) A distribution is said to have *Generalized AGG* right tail, if $\log \overline{F}$ is regularly varying,

$$\log \overline{F}(x) = -x^\nu L(x), \qquad (2.42)$$

where $\nu > 0$ and $L : (0, +\infty) \to (0, +\infty)$ is a slowly varying function. (A function is said to be slowly varying if $\lim_{x \to \infty} L(tx)/L(x) = 1$ for all $t > 0$.) Note that the AGG(ν) model corresponds to the special case where $L(x) \to 1/\nu$, as $x \to \infty$.

Relation (2.39) holds for all arrays $\mathcal{E}$ with *generalized* AGG marginals; if the entries are independent, the maxima are relatively stable. This follows directly from Proposition 2.2, once we show that F has rapidly varying tail. Indeed, by (2.42), we have

$$\log\left(\overline{F}(tx)\Big/\overline{F}(x)\right) = -L(x)x^\nu\left(t^\nu\frac{L(tx)}{L(x)} - 1\right),$$

which converges to $-\infty$, 0, and $+\infty$, as $x \to \infty$, when $t > 1$, $t = 1$, and $t < 1$, respectively, since $x^\nu L(x) \to \infty$ as $x \to \infty$ by definition of L.

The AGG class encompasses a wide variety of rapidly varying tail models such as Gaussian and double exponential distributions. The larger class (2.42) is needed, however, for the Gamma distribution.

More generally, distributions with heavier tails (e.g., log-normal) and lighter tails (e.g., Gompertz) outside the generalized AGG class (2.42) may also possess rapidly varying tails; heavy-tailed distributions like the Pareto and t-distributions, on the other hand, do not. More details on these alternative classes of models can be found in Appendix B.

2.7 Auxiliary Facts About Gaussian Distributions

We end this chapter with several facts about univariate and multivariate Gaussian distributions that will be used in the rest of the manuscript.

Relative stability. We first state the relative stability of iid standard Gaussian random variables. Since the standard Gaussian distribution falls in the class of asymptotically generalized Gaussians (AGG; see Definition 2.6), by Example 2.1, we know that the triangular array $\mathcal{E} = \{(\epsilon_p(i))_{i=1}^p, p \in \mathbb{N}\}$ has relatively stable (RS) maxima in the sense of (2.38), i.e.,

$$\frac{1}{u_p} \max_{i=1,\dots,p} \epsilon_p(i) \xrightarrow{\mathbb{P}} 1, \quad \text{as } p \to \infty, \tag{2.43}$$

where u_p is the $(1/p)$-th upper quantile as defined in (2.33). Similarly, since the array $\mathcal{E}$ has distributions symmetric around 0, it also has relatively stable minima

$$\frac{1}{u_p} \min_{i=1,\dots,p} \epsilon_p(i) \xrightarrow{\mathbb{P}} -1, \quad \text{as } p \to \infty. \tag{2.44}$$

The convergence in (2.43) also holds almost surely.

Mill's ratio. We give next the well-known bounds for Mill's ratio of Gaussian tails. Let Φ denote the CDF of the standard Gaussian distribution and ϕ its density. One can show that for all $x > 0$ we have

$$\frac{x}{1+x^2}\phi(x) \le \overline{\Phi}(x) = 1 - \Phi(x) \le \frac{1}{x}\phi(x), \tag{2.45}$$

using, e.g., integration by parts. Note that this fact may be used to verify the rapid variation of Φ, which entails the relative stability property above.

Stochastic monotonicity. The third fact is the stochastic monotonicity of the Gaussian location family. In fact, for all location families $\{F_\delta(x)\}_\delta$ where $F_\delta(x) = F(x - \delta)$, we have

$$F_{\delta_1}(t) \ge F_{\delta_2}(t), \quad \text{for all } t \in \mathbb{R} \text{ and all } \delta_1 \le \delta_2. \tag{2.46}$$

Relation (2.46) holds, of course, when F is the standard Gaussian distribution.

Slepian's lemma and the Sudakov–Fernique inequality. The following two results will be instrumental in our characterization of uniform relative stability for Gaussian triangular arrays in Chap. 6. The first is the celebrated result due to Slepian (1962).

Theorem 2.1 (Slepian's lemma) *Let* $\epsilon = (\epsilon(i))_{i=1}^{p}$ *and* $\eta = (\eta(i))_{i=1}^{p}$ *be two multivariate normally distributed random vectors with zero means* $\mathbb{E}[\epsilon(i)] = \mathbb{E}[\eta(i)] = 0$.
If for all $i, j = 1, \cdots, p$, *we have*

$$\mathbb{E}[\epsilon(i)^2] = \mathbb{E}[\eta(i)^2], \quad and \quad Cov(\epsilon(i), \epsilon(j)) \leq Cov(\eta(i), \eta(j)),$$

then $\epsilon \overset{st}{\geq} \eta$, *i.e.,*

$$\mathbb{P}[\epsilon(i) \leq x_i, \ i = 1, \cdots, p] \leq \mathbb{P}[\eta(i) \leq x_i, \ i = 1, \cdots, p].$$

This result implies, in particular, that $M_\epsilon := \max_{i=1,\cdots,p} \epsilon(i)$ dominates stochastically $M_\eta := \max_{i=1,\cdots,p} \eta(i)$ in the sense that

$$\mathbb{P}[M_\eta > u] \leq \mathbb{P}[M_\epsilon > u], \quad \text{for all } u \in \mathbb{R}. \tag{2.47}$$

In this case, we shall write $M_\eta \overset{d}{\leq} M_\epsilon$. This result shows, for example, that the maximum of iid Gaussians is stochastically larger than the maximum of any positively correlated Gaussian vector with the same marginal distributions.

Slepian's lemma can be obtained as a corollary from the general Normal Comparison Lemma (see, e.g., Theorem 4.2.1 on page 81 in Leadbetter et al. 1983). See also Chap. 2 in Adler and Taylor (2009).

The following result, known as the Sudakov–Fernique inequality, is similar in spirit to Slepian's lemma but it does not assume that the Gaussian vectors are centered and yield a weaker conclusion—an inequality between expectations. For a proof, many insights, and, in fact, a more general result, see, e.g., Theorem 2.2.5 on page 61 in Adler and Taylor (2009).

Theorem 2.2 (The Sudakov–Fernique inequality) *Let* $\epsilon = (\epsilon(i))_{i=1}^{p}$ *and* $\eta = (\eta(i))_{i=1}^{p}$ *be two multivariate normally distributed random vectors.*
If for all $i, j = 1, \cdots, p$, *we have*

$$\mathbb{E}[\epsilon(i)] = \mathbb{E}[\eta(i)] \quad and \quad \mathbb{E}[(\eta(i) - \eta(j))^2] \leq \mathbb{E}[(\epsilon(i) - \epsilon(j))^2],$$

then for $M_\epsilon = \max_{i=1,\cdots,p} \epsilon(i)$ *and* $M_\eta = \max_{i=1,\cdots,p} \eta(i)$, *we have*

$$\mathbb{E}[M_\eta] \leq \mathbb{E}[M_\epsilon].$$

With these conceptual and technical preparations, we are ready to discuss the high-dimensional signal detection and support recovery problems in the next chapter.

Chapter 3
A Panorama of Phase Transitions

The purpose of this chapter is to provide a unified review of the fundamental statistical limits in the sparse signal detection and support recovery problems. Our goal is to convey the main ideas and thus we shall focus on the simple but important setting of independent Gaussian errors. Specifically, we derive the conditions under which the detection and support recovery problems succeed and fail in the sense of (2.25) and (2.26), in the additive error model

$$x(i) = \mu(i) + \epsilon(i), \quad i = 1, \dots, p, \tag{3.1}$$

where the errors $\epsilon(i)$'s are iid standard Gaussians random variables. Once again, we restrict our analysis to models with independent and identically distributed Gaussian errors for the moment. Both the distributional assumption and the independence assumption will be relaxed substantially in the following chapters.

As laid out in Sect. 2.3, we work under the asymptotic regime where the problem dimension p diverges to infinity. The set of non-zero entries of the signal vector $\mu = \mu_p$ will be referred to as its *support* and denoted by

$$S_p := \{i \ : \ \mu(i) \neq 0\}.$$

We shall assume that the size of the support is

$$|S_p| = \lfloor p^{1-\beta} \rfloor, \quad \beta \in (0, 1], \tag{3.2}$$

where β parametrizes the problem sparsity. A more general parameterization of the support involving a slowly varying function is considered in Chap. 4.

© The Author(s), under exclusive license to Springer Nature Switzerland AG 2021
Z. Gao and S. Stoev, *Concentration of Maxima and Fundamental Limits in High-Dimensional Testing and Inference*,
SpringerBriefs in Probability and Mathematical Statistics,
https://doi.org/10.1007/978-3-030-80964-5_3

The closer β to 1, the sparser the support S_p. Conversely, when β is close to 0, the support is dense with many non-null signals. We consider one-sided alternatives (2.14), and parametrize the range of the non-zero (and perhaps unequal) signals with

$$\underline{\Delta} = \sqrt{2\underline{r}\log p} \le \mu(i) \le \overline{\Delta} = \sqrt{2\overline{r}\log p}, \quad \text{for all } i \in S_p, \tag{3.3}$$

for some constants $0 < \underline{r} \le \overline{r} \le +\infty$.

The parametrization of signal sparsity (3.2) and signal sizes (3.3) in the Gaussian model was first introduced in Ingster (1998), and later adopted by Hall and Jin (2010), Cai et al. (2011), Zhong et al. (2013), Cai and Wu (2014), Arias-Castro and Wang (2017), and numerous others for studying the signal detection problem in Gaussian location-scale models. Similar scalings of sparsity and signal size are also used in, e.g., Ji and Jin (2012), Jin et al. (2014), Butucea et al. (2018) to study the phase transitions of the support recovery problems under Gaussianity assumptions.

It should be noted that the "classical" setting where all signals are of equal size is not the only one that have been studied. The recent contribution of Li and Fithian (2020) investigates the signal detection problem in a more realistic setting where the signals are drawn from a general and potentially polynomial-tailed distribution. The study of such general settings in both detection and support recovery problems is an interesting new direction of research.

3.1   Sparse Signal Detection Problems

The optimality of sparse signal detection was first studied by Ingster (1998), who showed that a phase transition in the r-β plane exists for the signal detection problem. Specifically, consider the so-called *detection boundary* function:

$$f_D(\beta) = \begin{cases} \max\{0, \beta - 1/2\} & 0 < \beta \le 3/4 \\ \left(1 - \sqrt{1-\beta}\right)^2 & 3/4 < \beta \le 1. \end{cases} \quad \beta \in (0, 1]. \tag{3.4}$$

Assume that the non-zero signal sizes are all equal and parameterized as $\sqrt{2r\log p}$. If the signal size parameter r is *above* the detection boundary, i.e., $r > f_D(\beta)$, then the global null hypothesis $\mu(i) = 0$ for all $i = 1, \ldots, p$ can be distinguished from the alternative as $p \to \infty$ in the sense of (2.25) using the likelihood ratio test. Otherwise, when the signal sizes fall below the boundary, i.e., $r < f_D(\beta)$, no test can do better than a random guess. We visualize the detection boundary in the upper panel of Fig. 3.1.

Adaptive tests such as Tukey's HC in (2.19) (Donoho and Jin 2004) and a modified goodness-of-fit test statistic of Zhang (2002) have been identified to attain this performance limit without knowledge of the sparsity and signal sizes. It is also known that the max-statistic (2.16) is only efficient when $r > (1 + \sqrt{1-\beta})^2$, and is therefore sub-optimal for denser signals where $1/2 \le \beta \le 3/4$; see Cai et al. (2011).

(Recently, Li and Fithian 2020 showed that in the more general setting where signals themselves are dispersed, the sub-optimality of the max-statistic disappears in the detection problem.) In contrast, the sum-of-square-type statistics such as L_2 was shown in Fan (1996) to be asymptotically powerless when the L_2-norm of the signal $\|\mu\|_2^2$ is $o(\sqrt{p})$ or, equivalently, when $\beta > 1/2$ in our parametrization.

Notice that the scaling for the signal magnitude $\Delta = \sqrt{2r \log p}$ is useful for studying very sparse signals ($\beta > 1/2$), but fails to reveal the difficulties of the detection problems when the signals are relatively dense ($\beta < 1/2$). This is because $f_D(\beta) = 0$, $\beta \in (0, 1/2]$. Thus, a different scaling is needed to study the regime of small but dense signals. In this case, with slight overloading of notation, we parametrize signal sizes as

$$\underline{\Delta} = p^{\underline{r}} \leq \mu(i) \leq \overline{\Delta} = p^{\overline{r}}, \quad \text{for all } i \in S_p, \tag{3.5}$$

where $\underline{r}$ and $\overline{r}$ are negative constants and the signal magnitude vanishes, as $p \to \infty$. In this scaling, for the so-called faint signal regime, Cai et al. (2011) established a phase-transition result characterized by the following boundary:

$$f_{D'}(\beta) = \beta - 1/2, \quad 0 < \beta \leq 1/2. \tag{3.6}$$

Specifically, if $\overline{r} < f_{D'}(\beta)$, the signal detection fails in the sense of (2.26) regardless of the procedures, while the HC statistic continues to attain asymptotically perfect detection when $\underline{r} > f_{D'}(\beta)$. We visualize this boundary in the lower panel of Fig. 3.1.

To the best of our knowledge, the performance of simple statistics such as L_1, L_2 norms, and the sum statistic T in (2.15) in this weak signal setting have not been reported in the literature. Our first theorem establishes the performance of these simple but popular statistics for detecting sparse signals in high dimensions, and summarizes the known results.

Theorem 3.1 *Consider the signal detection problem in the triangular array of Gaussian error models* (3.1) *where the sparsity is parametrized as in* (3.2).

(i) *For $\beta \in (1/2, 1)$ and growing signal sizes as in* (3.3), *the statistics L_1, L_2, and T are asymptotically powerless in the sense of* (2.26).

(ii) *For $\beta \in (0, 1/2]$ and growing signal sizes as in* (3.3), *the statistics L_1, L_2, and T solve the detection problem in the sense of* (2.25).

(iii) *For dense and faint signals, i.e., $\beta \in (0, 1/2]$ under the parameterization* (3.5), *the sum statistic T attains the optimal detectability boundary in* (3.6). *That is, tests based on the sum statistic T can succeed asymptotically in the sense of* (2.25) *when $\underline{r} > \beta - 1/2$.*

(iv) *In the dense and faint signal setting of* (iii), *the L_1 and L_2 statistics are both sub-optimal. More precisely, they succeed in the sense of* (2.25) *when $\underline{r} > \beta/2 - 1/4$, but fail in the sense of* (2.26) *when $\overline{r} < \beta/2 - 1/4$.*

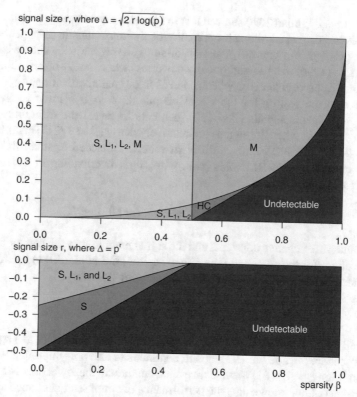

Fig. 3.1 The phase diagrams of the sparse signal detection problem. Signal size and sparsity are parametrized by r and β, respectively. The diagrams illustrate the regions where the signal detection problem can be solved asymptotically by some of the commonly used statistics: the maximum (M), the sum-of-squares (L_2), the sum-of-absolute values (L_1), and the sum (T). In each region of the diagram, the annotated statistics can make the detection risk (2.4) vanish, as dimension p diverges. Conversely, the risks has liminf at least one. The detection problem is unsolvable for very sparse and weak signals in the undetectable regions. Notice that the L_1 and L_2 statistics are, in fact, sub-optimal for all sparsity levels. On the other hand, the max-statistic remains powerful for sparse signals ($\beta > 1/2$), and is fully efficient when the problem is very sparse ($\beta \geq 3/4$). The HC statistic can detect signals in all configurations in the detectable regions; we explicitly marked the region where signals are only detectable by HC among the statistics considered. See the text and Theorem 3.1

Proof The claims in parts (i) and (ii) about the L_1, L_2, and the sum statistic T in the cases of diverging signal sizes (3.3) can be found in Fan (1996), Candès (2018). We prove here the statements for the cases (iii) and (iv), where the signals are dense and small, as parametrized in (3.5).

For simplicity of the exposition, we will suppose that in (3.5) we have $\underline{r} = r = \overline{r}$, so that $\mu(i) = p^r$. The general case where $\underline{r} < \overline{r}$ is left as an exercise.

Part (iii): We first show that the sum statistic T or, equivalently, the simple arithmetic mean attains the sparse signal detection boundary. By the normality and independence of the summands, we have

$$\frac{1}{\sqrt{p}} \sum_{i=1}^{p} x(i) \sim \begin{cases} N(0, 1), & \text{under } H_0 \\ N(p^{(r-\beta)+1/2}, 1), & \text{under } H_1. \end{cases} \tag{3.7}$$

It immediately follows that the two distributions can be distinguished perfectly if $p^{r-(\beta-1/2)}$ diverges, i.e., $r > \beta - 1/2$. This can be seen by simply setting the rejection region at $(p^{(r-\beta)+1/2}/2, +\infty)$ for the scaled statistic $\sum_{i=1}^{p} x(i)/\sqrt{p}$. According to the lower bound on the performance limit in detection problems (see Theorem 8 in Cai et al. 2011), we have shown that T attains the optimal detection boundary (3.6).

Part (iv): We now turn to the L_2-norm statistic. Recall a non-central chi-square random variable $\chi_k^2(\lambda)$ has mean $(k + \lambda)$ and variance $2(k + 2\lambda)$. Since the observations have distributions $N(0, 1)$ under the null and $N(p^r, 1)$ under the alternative, we have $x^2(i) \sim \chi_1^2(0)$ for $i \notin S$ and $x^2(i) \sim \chi_1^2(p^{2r})$ for $i \in S$. Therefore, the mean and variance of the (centered and scaled) L_2 statistics are

$$\mathbb{E}\left[\frac{1}{\sqrt{p}} \sum_{i=1}^{p} \left(x(i)^2 - 1\right)\right] = \begin{cases} 0 & \text{under } H_0 \\ p^{1-\beta} p^{2r} p^{-1/2} = p^{1/2-\beta+2r} & \text{under } H_1, \end{cases} \tag{3.8}$$

and

$$\text{Var}\left(\frac{1}{\sqrt{p}} \sum_{i=1}^{p} \left(x(i)^2 - 1\right)\right) = \begin{cases} \frac{1}{p} 2p = 2 & \text{under } H_0 \\ \frac{1}{p} \left(2p + 4p^{1-\beta+2r}\right) = 2(1 + 2p^{2r-\beta}) & \text{under } H_1, \end{cases} \tag{3.9}$$

respectively. By the central limit theorem, we have

$$\frac{1}{\sqrt{2p}} \sum_{i=1}^{p} \left(x(i)^2 - 1\right) \implies N(0, 1), \tag{3.10}$$

under the null. On the other hand, under the alternative, since $p^{2r-\beta} \to 0$ for all $r < 0$ and $\beta > 0$, the variance in (3.9) converges to 2, as $p \to \infty$ and an application of the Lyapunov version of the CLT entails

$$\frac{1}{\sqrt{2p}} \left(\sum_{i=1}^{p} \left(x(i)^2 - 1\right) - p^{1/2-\beta+2r}\right) \implies N(0, 1). \tag{3.11}$$

Hence, perfect detection with the L_2-norm is possible if $p^{1/2-\beta+2r}$ diverges, i.e., $r > \beta/2 - 1/4$. On the other hand, if $r < \beta/2 - 1/4$, the distributions of the (scaled) statistics merge under the null and the alternative.

The case of the L_1-norm is treated similarly. Let $Y = |X|$ where $X \sim |N(\mu, 1)|$. Using the expressions for the mean and variance of Y (see, e.g., Tsagris et al. 2014),

$$\mu_Y = \mathbb{E}[Y] = \sqrt{\frac{2}{\pi}} e^{-\mu^2/2} + \mu(1 - \Phi(-\mu)), \tag{3.12}$$

$$\sigma_Y^2 = \mathrm{Var}(Y) = \mu^2 + 1 - \mu_Y^2, \tag{3.13}$$

where Φ is the CDF of a standard normal random variable, we have, regardless of the value of μ,

$$\sigma_Y^2 = \mathrm{Var}(Y) = \mathbb{E}(Y - \mathbb{E}Y)^2 \leq \mathbb{E}(X - \mathbb{E}X)^2 = 1, \tag{3.14}$$

where the inequality holds because absolute value is a Lipschitz function with Lipschitz constant 1.

By the central limit theorem, we have

$$\frac{1}{\sqrt{p}} \left(\sum_{i=1}^p |x(i)| - \sqrt{\frac{2}{\pi}} \right) \implies N(0, 1 - 2/\pi) \tag{3.15}$$

under the null. On the other hand, when the alternative hypothesis holds, we have

$$\mathbb{E}\left[\frac{1}{\sqrt{p}} \left(\sum_{i=1}^p |x(i)| - \sqrt{\frac{2}{\pi}} \right) \right] = \frac{p^{1-\beta}}{\sqrt{p}} \left[\left(\sqrt{\frac{2}{\pi}} e^{-\mu^2/2} + \mu\,(1 - 2\Phi(-\mu)) \right) - \sqrt{\frac{2}{\pi}} \right]$$

$$= p^{1/2-\beta} \left[\sqrt{\frac{2}{\pi}} \left(e^{-p^{2r}/2} - 1 \right) + p^r \left(1 - 2\Phi(-\mu) \right) \right]$$

$$= p^{1/2-\beta} \left[\sqrt{\frac{2}{\pi}} \left(-p^{2r}/2 - O(p^{4r}) \right) + p^r \left(\sqrt{\frac{2}{\pi}} p^r + O(p^{3r}) \right) \right]$$

$$= p^{1/2-\beta} \sqrt{\frac{2}{\pi}} \left(p^{2r}/2 + O(p^{4r}) \right)$$

$$= p^{1/2-\beta+2r} \sqrt{1/2\pi} + O(p^{1/2-\beta+4r}),$$

and

$$\mathrm{Var}\left(\frac{1}{\sqrt{p}} \left(\sum_{i=1}^p |x(i)| - \sqrt{\frac{2}{\pi}} \right) \right) = \frac{1}{p}(p - p^{1-\beta})(1 - 2/\pi) + \frac{1}{p} p^{1-\beta} \sigma_Y^2$$

$$\to 1 - 2/\pi,$$

by the boundedness of σ_Y^2 shown in (3.14). Again, by the Lyapunov version of the central limit theorem, we conclude asymptotic normality of the centered and scaled L_1-norms under the alternative. In an entirely analogous argument to the L_2-norm case, asymptotically perfect detection can be achieved if $p^{1/2-\beta+2r}$ diverges, i.e., $r > \beta/2 - 1/4$. On the other hand, when $r < \beta/2 - 1/4$, the two hypotheses cannot be told apart by the L_1-norms since the distributions of the (scaled) statistics merge under the two hypotheses. $\qquad\square$

The portmanteau of results in Theorem 3.1 are visualized in Fig. 3.1. It is worth noting that the β-r parameter regions where L_1 and L_2 statistics are asymptotically powerful coincide, and these statistics are theoretically sub-optimal for both sparse regimes ($\beta > 1/2$) and relatively dense regimes ($\beta \leq 1/2$).

Ideas have been proposed to combine statistics that are powerful for different alternatives to create adaptive tests that maintain high power for all sparsity levels. Such adaptive tests can be constructed, for example, by leveraging the asymptotic independence of the sum- and supremum-type statistics (Hsing 1995). Recently, Xu et al. (2016) showed that for dependent observations under mixing and moment conditions, the sum-of-power-type statistics

$$\tilde{L}_q(x) = \sum_{i=1}^{p} x^q(i) \tag{3.16}$$

with distinct positive integer powers (i.e., $q = 1, 2, \ldots$) are asymptotically jointly independent, and proposed an adaptive test that monitors the minimum p-value of tests constructed with $\tilde{L}_q$'s. This idea is further developed in Wu et al. (2019) for generalized linear models and in He et al. (2018) with U-statistics.

Optimality properties of such adaptive tests and the optimal choice of the q-combinations, however, remain open problems. Xu et al. (2016) suggested combining $q = 1, 2, 3, \ldots, 6$, and $q = \infty$, based empirical evidence from numerical experiments. Theorem 3.1 here implies that, at least for detecting one-sided alternatives, the $\tilde{L}_2$ statistic (i.e., L_2 norm) and the L_1 norm are asymptotically dominated by the $\tilde{L}_1$ statistic (or, equivalently, the sum T). Therefore, it is sufficient to include only the latter in the construction of the adaptive test.

3.2 Sparse Signal Support Recovery Problems

Turning to support recovery problems in the Gaussian error model (3.1), in the rest of this chapter, we will analyze the asymptotic performance limits in terms of the risk metrics for exact, exact–approximate, approximate–exact support recovery problems (i.e., (2.8), (2.11), and (2.12), respectively), as well as the probability of support recovery (2.9). We will also review the recent result for exact support recovery risk (2.7) by Arias-Castro and Chen (2017) to reveal a rather complete landscape of support recovery problems in high-dimensional Gaussian error models.

In the rest of this chapter, we restrict our attention to the class of thresholding procedures. Specifically, the lower bounds that we develop in Theorems 3.2 through 3.5 are only meant to apply to thresholding procedures. Although it is intuitively appealing to consider only data-thresholding procedures in multiple testing problems, such procedures are not always optimal in more general settings. The optimality of thresholding procedures and the consequences of this restriction will be treated in Chap. 5.

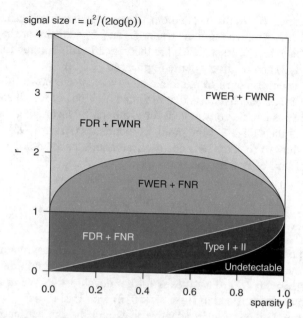

Fig. 3.2 The phase diagram of support recovery problems for the high-dimensional model (3.1), illustrating the boundaries of the exact support recovery (FWER + FWNR; top curve; Theorem 3.2), the approximate-exact support recovery (FDR + FWNR; second curve from top; Theorem 3.5), the exact–approximate support recovery (FWER + FNR; horizontal line $r = 1$; Theorem 3.4), and the approximate support recovery problems (FDR + FNR; tilted line $r = \beta$; Theorem 3.3). The signal detection problem (Type I + Type II errors of the global test; lower curve) was studied in Donoho and Jin (2004). In each region of the diagram and above, the annotated statistical risk can be made to vanish, as dimension p diverges. Conversely, the risks have liminf at least one

Figure 3.2 illustrates the rich landscape of phase transitions in support recovery for the various choices of statistical risk for the family of thresholding estimators, established in the following sections. We end this brief overview with a technical notion needed in order to state our main results. We define a rate at which the nominal levels of FWER or FDR go to zero.

Definition 3.1 We say the nominal level of errors $\alpha = \alpha_p$ vanishes slowly, if

$$\alpha \to 0, \quad \text{and} \quad \alpha p^\delta \to \infty \text{ for any } \delta > 0. \tag{3.17}$$

As an example, the sequence of nominal levels $\alpha_p = 1/\log(p)$ is slowly vanishing, while the sequence $\alpha_p = 1/\sqrt{p}$ is not.

3.3 The Exact Support Recovery Problem

Our study of the exact support recovery risk (2.8) begins with a brief review of existing results for the Hamming loss (2.10). Indeed, as discussions in Sect. 2.3 suggest, the latter can be informative of the exact support recovery problems for models with independent components.

Inspired by the phase-transition results for the signal detection problem, Ji and Jin (2012), Genovese et al. (2012), Jin et al. (2014) derived interesting sharp results on support recovery problems in linear models under the Hamming loss $H(\widehat{S}, S)$. Specifically, these papers establish minimax-type phase-transition results in their respective settings. Under the sparsity parametrization in (3.2) and assuming equal signal sizes of $(2r \log p)^{1/2}$, Hamming losses were shown to diverge to $+\infty$ when r falls below the threshold

$$f_{\mathrm{E}}(\beta) = (1 + (1 - \beta)^{1/2})^2, \qquad (3.18)$$

for any method of support estimation. Conversely, under orthogonal, or near-orthogonal random designs, if $r > f_{\mathrm{E}}(\beta)$, they showed that the methods they proposed achieve vanishing Hamming loss.

Very recently, Butucea et al. (2018) studied both asymptotics and non-asymptotics of support recovery problems in the additive noise model (3.1) under the assumption of equal signal sizes, using the Hamming loss. Again, the analysis of asymptotic optimality focused on a newly proposed procedure which is very specific to the Gaussian model. It is not at all clear if the optimality properties are a consequence of its mysterious construction.

We now show that commonly used and computationally efficient procedures can also be asymptotically optimal in the exact support recovery problem.

Theorem 3.2 *Consider the high-dimensional additive error model* (3.1) *under independent standard Gaussian errors, with signal sparsity and size as described in* (3.2) *and* (3.3). *The function* (3.18) *characterizes the phase transition of the exact support recovery problem. Namely, the following two results hold:*

(i) *If $\underline{r} > f_{\mathrm{E}}(\beta)$, then Bonferroni's, Sidák's, Holm's, and Hochberg's procedures with slowly vanishing nominal FWER levels (as defined in Definition 3.1) all achieve asymptotically exact support recovery in the sense of* (2.25).
(ii) *Conversely, if $\overline{r} < f_{\mathrm{E}}(\beta)$, then for any thresholding procedure $\widehat{S}_p$, we have $\mathbb{P}[\widehat{S}_p = S_p] \to 0$. Therefore, in view of Lemma 2.1, exact support recovery asymptotically fails for all thresholding procedures in the sense of* (2.26).

We illustrate this result with a β-r phase diagram in Fig. 3.2. Theorem 3.2 is, in fact, a special case of the more general Theorem 4.1, which covers dependent as well as Gaussian and non-Gaussian errors. We will study the *exact support recovery* problem in greater detail and generality in Chap. 4.

3.4   The Approximate Support Recovery Problem

Arias-Castro and Chen (2017) studied the performance of the Benjamini–Hochberg procedure (Benjamini and Hochberg 1995) and a stripped-down version of the Candés–Barber procedure (Barber and Candès 2015) in approximate support recovery problems when the components of the noise term ϵ in (3.1) have independent and symmetric distributions. A phase-transition phenomenon for the approximate support recovery risk (2.7) was established in the Gaussian additive error model, where the two aforementioned methods are both shown to be asymptotically optimal.

The analysis therein, however, assumed equal signal sizes for the alternatives. We generalize the main results of Arias-Castro and Chen (2017) to allow for unequal signal sizes. The key to establishing this generalization is a monotonicity property of the BH procedure, presented in the following Sect. 3.5. Namely, the power of the BH procedure in terms of FNR monotonically increases for stochastically larger alternatives. This fact will be formalized in Lemma 3.2, and may be of independent interest.

Theorem 3.3 *In the context of Theorem 3.2, the function*

$$f_A(\beta) = \beta \tag{3.19}$$

characterizes the phase transition of the approximate support recovery problem. Specifically the following two results hold:

(i) *If $\underline{r} > f_a(\beta)$, then the Benjamini–Hochberg procedure (defined in Sect. 2.2) with slowly vanishing nominal FDR levels (as defined in Definition 3.1) achieves asymptotically approximate support recovery in the sense of (2.25).*
(ii) *Conversely, if $\overline{r} < f_a(\beta)$, then approximate support recovery asymptotically fails in the sense of (2.26) for all thresholding procedures.*

Proof (*Necessary condition in Theorem 3.3*) We first show part (ii). That is, when $\overline{r} < \beta$, no thresholding procedure is able to achieve approximate support recovery. The arguments are similar to those in the proof of Theorem 1 of Arias-Castro and Chen (2017), although we allow for unequal signal sizes.

Denote the distributions of $N(0, 1)$, $N(\underline{\Delta}, 1)$, and $N(\overline{\Delta}, 1)$ as F_0, $F_{\underline{a}}$, and $F_{\overline{a}}$, respectively.

Recall that thresholding procedures are of the form

$$\widehat{S}_p = \left\{ i \mid x(i) > t_p(x) \right\}.$$

Denote $\widehat{S} := \left\{ i \mid x(i) > t_p(x) \right\}$, and $\widehat{S}(u) := \{ i \mid x(i) > u \}$. For any threshold $u \geq t_p$, we must have $\widehat{S}(u) \subseteq \widehat{S}$, and hence

$$\text{FDP} := \frac{|\widehat{S} \setminus S|}{|\widehat{S}|} \geq \frac{|\widehat{S} \setminus S|}{|\widehat{S} \cup S|} = \frac{|\widehat{S} \setminus S|}{|\widehat{S} \setminus S| + |S|} \geq \frac{|\widehat{S}(u) \setminus S|}{|\widehat{S}(u) \setminus S| + |S|}. \tag{3.20}$$

On the other hand, for any threshold $u \leq t_p$ we must have $\widehat{S}(u) \supseteq \widehat{S}$, and hence

$$\text{NDP} := \frac{|S \setminus \widehat{S}|}{|S|} \geq \frac{|S \setminus \widehat{S}(u)|}{|S|}. \tag{3.21}$$

Since either $u \geq t_p$ or $u \leq t_p$ must take place, putting (3.20) and (3.21) together, we have

$$\text{FDP} + \text{NDP} \geq \frac{|\widehat{S}(u) \setminus S|}{|\widehat{S}(u) \setminus S| + |S|} \wedge \frac{|S \setminus \widehat{S}(u)|}{|S|}, \tag{3.22}$$

for any u. Therefore, it suffices to show that for a suitable choice of u, the RHS of (3.22) converges to 1 in probability; the desired conclusion on FDR and FNR follows by the dominated convergence theorem.

Let $t^* = \sqrt{2q \log p}$ for some fixed q, we obtain an estimate of the tail probability by Mill's ratio (2.45),

$$\overline{F_0}(t^*) \sim \frac{1}{t^*} \phi(t^*) = \frac{1}{2\sqrt{\pi q \log p}} p^{-q}, \tag{3.23}$$

where $a_p \sim b_p$ is taken to mean $a_p / b_p \to 1$. Observe that $|\widehat{S}(t^*) \setminus S|$ has distribution $\text{Binom}(p - s, \overline{F_0}(t^*))$ where $s = |S|$, denote $X = X_p := |\widehat{S}(t^*) \setminus S|/|S|$, and we have

$$\mu := \mathbb{E}[X] = \frac{(p - s)\overline{F_0}(t^*)}{s}, \quad \text{and} \quad \text{Var}(X) = \frac{(p - s)\overline{F_0}(t^*)F_0(t^*)}{s^2} \leq \mu/s.$$

Therefore, for any $M > 0$, we have, by Chebyshev's inequality,

$$\mathbb{P}[X < M] \leq \mathbb{P}[|X - \mu| > \mu - M] \leq \frac{\mu/s}{(\mu - M)^2} = \frac{1/(\mu s)}{(1 - M/\mu)^2}. \tag{3.24}$$

Now, from the expression of $\overline{F_0}(t^*)$ in (3.23), we obtain

$$\mu = (p^\beta - 1)\overline{F_0}(t^*) \sim \frac{1}{2\sqrt{\pi q \log p}} p^{\beta - q}.$$

Since $\bar{r} < \beta$, we can pick q such that $\bar{r} < q < \beta$. In turn, we have $\mu \to \infty$, as $p \to \infty$. Therefore, the last expression in (3.24) converges to 0, and we conclude that $X \to \infty$ in probability, and hence

$$\frac{|\widehat{S}(t^*) \setminus S|}{|\widehat{S}(t^*) \setminus S| + |S|} = \frac{X}{X + 1} \to 1 \quad \text{in probability.} \tag{3.25}$$

On the other hand, we show that with the same choice of $u = t^*$, we have

$$\frac{|S \setminus \widehat{S}(t^*)|}{|S|} \to 1 \quad \text{in probability.} \tag{3.26}$$

By the stochastic monotonicity of Gaussian location family (2.46), we have the following lower bound for the probability of missed detection for each signal $\mu(i)$, $i \in S$,

$$\mathbb{P}[N(\mu(i), 1) \le t^*] \ge F_{\bar{a}}(t^*). \tag{3.27}$$

Since $|S \setminus \widehat{S}(t^*)|$ can be written as the sum of s independent Bernoulli random variables,

$$|S \setminus \widehat{S}(t^*)| = \sum_{i \in S} \mathbb{1}_{(-\infty, t^*]}(x(i)),$$

using (3.27), we conclude that $|S \setminus \widehat{S}(t^*)| \overset{d}{\ge} \mathrm{Binom}(s, F_{\bar{a}}(t^*))$. Finally, we know that $F_{\bar{a}}(t^*)$ converges to 1 by our choice of diverging t^*, and the necessary condition is shown. $\qquad \square$

Proof *(Sufficient condition in Theorem 3.3)* We now turn to the sufficient condition, i.e., part (i). That is, when $\underline{r} > \beta$, the Benjamini–Hochberg procedure with slowly vanishing FDR levels achieves asymptotic approximate support recovery.

The FDR vanishes by our choice of α and the FDR-controlling property of the BH procedure (Benjamini and Hochberg 1995). It only remains to show that FNR also vanishes.

To do so we compare the FNR under the alternative specified in Theorem 3.3 to one with all of the signal sizes equal to $\underline{\Delta}$. By Lemma 3.2, it suffices to show that the FNR under the BH procedure in this setting vanishes. Let $x(i)$ be vectors of independent observations with $p - s$ nulls having standard Gaussian distributions, and s signals having $N(\underline{\Delta}, 1)$ distributions.

Denote the null and the alternative distributions as F_0 and F_a, respectively. Let $\widehat{G}$ denote the empirical survival function as in (3.36). Define the empirical survival functions for the null part and signal part

$$\widehat{W}_{\text{null}}(t) = \frac{1}{p-s} \sum_{i \notin S} \mathbb{1}\{x(i) \ge t\}, \quad \widehat{W}_{\text{signal}}(t) = \frac{1}{s} \sum_{i \in S} \mathbb{1}\{x(i) \ge t\}, \tag{3.28}$$

where $s = |S|$, so that

$$\widehat{G}(t) = \frac{p-s}{p} \widehat{W}_{\text{null}}(t) + \frac{s}{p} \widehat{W}_{\text{signal}}(t).$$

We need the following result to describe the deviations of the empirical distributions.

Lemma 3.1 (Theorem 1 of Eicker 1979) *Let* $Z_1, \ldots, Z_k$ *be iid with continuous survival function* Q. *Let* $\widehat{Q}_k$ *denote their empirical survival function and define* $\xi_k = \sqrt{2 \log \log (k)/k}$ *for* $k \ge 3$. *Then*

$$\frac{1}{\xi_k} \sup_z \frac{|\widehat{Q}_k(z) - Q(z)|}{\sqrt{Q(z)(1 - Q(z))}} \to 1,$$

in probability as $k \to \infty$. In particular,

$$\widehat{Q}_k(z) = Q(z) + O_{\mathbb{P}}\left(\xi_k \sqrt{Q(z)(1 - Q(z))}\right),$$

uniformly in z.

Applying Lemma 3.1 to the two summands in $\widehat{G}$, we obtain $\widehat{G}(t) = G(t) + \widehat{R}(t)$, where

$$G(t) = \frac{p - s}{p}\overline{F_0}(t) + \frac{s}{p}\overline{F_a}(t), \tag{3.29}$$

and

$$\widehat{R}(t) = O_{\mathbb{P}}\left(\xi_p\sqrt{\overline{F_0}(t)F_0(t)} + \frac{s}{p}\xi_s\sqrt{\overline{F_a}(t)F_a(t)}\right), \tag{3.30}$$

uniformly in t.

Recall (see proof of Lemma 3.2) that the BH procedure is the thresholding procedure with threshold set at

$$\tau = \inf\{t \mid \overline{F_0}(t) \le \alpha\widehat{G}(t)\}. \tag{3.31}$$

The NDP may also be re-written as

$$\text{NDP} = \frac{|S \setminus \widehat{S}|}{|S|} = \frac{1}{s}\sum_{i \in S} \mathbb{1}\{x(i) < \tau\} = 1 - \widehat{W}_{\text{signal}}(\tau),$$

so that it suffices to show that

$$\widehat{W}_{\text{signal}}(\tau) \to 1 \tag{3.32}$$

in probability. Applying Lemma 3.1 to $\widehat{W}_{\text{signal}}$, we know that

$$\widehat{W}_{\text{signal}}(\tau) = \overline{F_a}(\tau) + O_{\mathbb{P}}\left(\xi_s\sqrt{\overline{F_a}(\tau)F_a(\tau)}\right) = \overline{F_a}(\tau) + o_{\mathbb{P}}(1).$$

So it suffices to show that $F_a(\tau) \to 0$ in probability. Now let $t^* = \sqrt{2q\log(p)}$ for some q such that $\beta < q < \underline{r}$. We have

$$F_a(t^*) = \Phi(t^* - \underline{\Delta}) = \Phi(\sqrt{2(q - \underline{r})\log p}) \to 0. \tag{3.33}$$

Hence, in order to show (3.32), it suffices to show

$$\mathbb{P}\left[\tau \le t^*\right] \to 1. \tag{3.34}$$

By (3.29), the mean of the empirical process $\widehat{G}$ evaluated at t^* is

$$G(t^*) = \frac{p-s}{p}\overline{F_0}(t^*) + \frac{s}{p}\overline{F_a}(t^*). \tag{3.35}$$

The first term, using Relation (3.23), is asymptotic to $p^{-q}L(p)$, where $L(p)$ is the logarithmic term in p. The second term, since $\overline{F_a}(t^*) \to 1$ by Relation (3.33), is asymptotic to $p^{-\beta}$. Therefore, $G(t^*) \sim p^{-q}L(p) + p^{-\beta} \sim p^{-\beta}$, since $p^{\beta-q}L(p) \to 0$ where $q > \beta$.

The fluctuation of the empirical process at t^*, by Relation (3.30), is

$$\widehat{R}(t^*) = O_{\mathbb{P}}\left(\xi_p\sqrt{\overline{F_0}(t^*)F_0(t^*)} + \frac{s}{p}\xi_s\sqrt{\overline{F_a}(t^*)F_a(t^*)}\right)$$

$$= O_{\mathbb{P}}\left(\xi_p\sqrt{\overline{F_0}(t^*)}\right) + o_{\mathbb{P}}\left(p^{-\beta}\right).$$

By (3.23) and the expression for ξ_p, the first term is $O_{\mathbb{P}}\left(p^{-(q+1)/2}L(p)\right)$ where $L(p)$ is a poly-logarithmic term in p. Since $\beta < \min\{q, 1\}$, we have $\beta < (q+1)/2$, and hence $\widehat{R}(t^*) = o_{\mathbb{P}}(p^{-\beta})$.

Putting the mean and the fluctuation of $\widehat{G}(t^*)$ together, we obtain

$$\widehat{G}(t^*) = G(t^*) + \widehat{R}(t^*) \sim_{\mathbb{P}} G(t^*) \sim p^{-\beta},$$

and therefore, together with (3.23), we have

$$\overline{F_0}(t^*)/\widehat{G}(t^*) = p^{\beta-q}L(p)(1 + o_{\mathbb{P}}(1)),$$

which is eventually smaller than the FDR level α by Assumption (3.17) and the fact that $\beta < q$. That is,

$$\mathbb{P}\left[\overline{F_0}(t^*)/\widehat{G}(t^*) < \alpha\right] \to 1.$$

By the definition of τ (recall (3.31)), this implies that $\tau \le t^*$ with probability tending to 1, and (3.34) is shown. The proof for the sufficient condition is complete. □

3.5 Monotonicity of the Benjamini–Hochberg Procedure

As promised in the previous section, we make a connection between the power of the BH procedure and the stochastic ordering of distributions under the alternative. This natural result seems new.

Lemma 3.2 (Monotonicity of the BH procedure) *Consider p independent observations $x(i)$, $i \in \{1, \ldots, p\}$, where the $(p - s)$ coordinates in the null part have common distribution F_0, and the remaining s signals have continuous alternative* *distributions F_j^i, $i \in S$, respectively. Compare the two alternatives $j \in \{1, 2\}$, where the distributions in Alternative 2 are stochastically larger than those in Alternative 1, i.e.,*

$$F_2^i(t) \leq F_1^i(t), \quad \text{for all } t \in \mathbb{R}, \text{ and for all } i \in S.$$

If the BH procedure is applied at the same nominal level of FDR, then the FNR of the BH procedure under Alternative 2 is bounded above by the FNR under Alternative 1. Further, the threshold of the BH procedure under Alternative 2 is stochastically smaller than that under Alternative 1.

Loosely put, the power of the BH procedure is monotone increasing with respect to the stochastic ordering of the alternatives, yet (the distribution of) the BH threshold is monotone decreasing in the distributions of the alternatives.

Proof (*Lemma* 3.2) We first re-express the BH procedure in a different form. Recall that on observing $x(i)$, $i \in \{1, \ldots, p\}$, the BH procedure is the thresholding procedure with threshold set at $x_{[i^*]}$, where $i^* := \max\{i \mid \overline{F_0}(x_{[i]}) \leq \alpha i/p\}$, and $x_{[1]} \geq \ldots \geq x_{[p]}$ are the order statistics.

Let $\widehat{G}$ denote the left-continuous empirical survival function

$$\widehat{G}(t) = \frac{1}{p} \sum_{i=1}^{p} \mathbb{1}\{x(i) \geq t\}. \tag{3.36}$$

By the definition, we know that $\widehat{G}(x_{[i]}) = i/p$ a.s. (by continuity of the alternatives). Therefore, by the definition of i^*, we have

$$\overline{F_0}(x_{[i]}) > \alpha \widehat{G}(x_{[i]}) = \alpha i/p \quad \text{for all } i > i^*.$$

Since $\widehat{G}$ is constant on $(x_{[i^*+1]}, x_{[i^*]}]$, the fact that $\overline{F_0}(x_{[i^*]}) \leq \alpha \widehat{G}(x_{[i^*]})$ and $\overline{F_0}(x_{[i^*+1]}) > \alpha \widehat{G}(x_{[i^*+1]})$ implies that $\alpha \widehat{G}$ and $\overline{F_0}$ must "intersect" on the interval by continuity of F_0. We denote this "intersection" as

$$\tau = \inf\{t \mid \overline{F_0}(t) \leq \alpha \widehat{G}(t)\}. \tag{3.37}$$

Note that τ cannot be equal to $x_{[i^*+1]}$ since $\overline{F_0}$ is càdlàg. Since there is no observation in $[\tau, x_{[i^*]})$, we can write the BH procedure as the thresholding procedure with threshold set at τ.

Now, denote the observations under Alternatives 1 and 2 as $x_1(i)$ and $x_2(i)$. Since $x_2(i)$ stochastically dominates $x_1(i)$ for all $i \in \{1, \ldots, p\}$, there exists a coupling $(\tilde{x}_1, \tilde{x}_2)$ of x_1 and x_2 such that $\tilde{x}_1(i) \leq \tilde{x}_2(i)$ almost surely for all i. We will replace $\tilde{x}_1$ and $\tilde{x}_2$ with x_1 and x_2 in what follows. Since we will compare the FNRs, i.e., expectations with respect to the marginals of x's in the last step, this replacement does not affect the conclusions. To simplify notation, we still write x_1 and x_2 in place of $\tilde{x}_1$ and $\tilde{x}_2$.

Let $\widehat{G}_k$ be the left-continuous empirical survival function under Alternative k, i.e.,

$$\widehat{G}_k(t) = \frac{1}{p} \sum_{i=1}^{p} \mathbb{1}\{x_k(i) \geq t\}, \quad k \in \{1, 2\}. \tag{3.38}$$

We define the BH thresholds τ_1 and τ_2 by replacing $\widehat{G}$ in (3.37) with $\widehat{G}_1$ and $\widehat{G}_2$, respectively. Denote the set estimates of signal support $\widehat{S}_k = \{i \mid x_k(i) \geq \tau_k\}$ by the BH procedure. We claim that

$$\tau_2 \leq \tau_1 \quad \text{with probability 1}. \tag{3.39}$$

Indeed, by definition of the empirical survival function (3.38) and the fact that $x_1(i) \leq x_2(i)$ almost surely for all i, we have $\widehat{G}_1(t) \leq \widehat{G}_2(t)$ for all t. Hence, $\overline{F}_0(t) \leq \alpha \widehat{G}_1(t)$ implies $\overline{F}_0(t) \leq \alpha \widehat{G}_2(t)$, and Relation (3.39) follows from the definition of τ in (3.37). The claim of stochastic ordering of the BH thresholds in Lemma 3.2 follows from (3.39).

Finally, when $\tau_2 \leq \tau_1$, we have $\tau_2 \leq \tau_1 \leq x_1(i) \leq x_2(i)$ with probability 1 for all $i \in \widehat{S}_1$. Therefore, it follows that $\widehat{S}_1 \subseteq \widehat{S}_2$ and hence $|S \setminus \widehat{S}_2| \leq |S \setminus \widehat{S}_1|$ almost surely. The first conclusion in Lemma 3.2 follows from the last inequality. □

3.6   The Exact–Approximate Support Recovery Problem

We now derive two new asymptotic phase-transition results for the *asymmetric* statistical risks, (2.11) and (2.12), in the Gaussian error models. As discussed in Sect. 2.1, the exact–approximate support recovery risk is the natural criteria when considering the marginal power of discovery while controlling for family-wise error rates in applications such as GWAS.

Although there have been discussions of weighted sums of Type I and Type II errors in the literature (see, e.g., Genovese and Wasserman (Genovese and Wasserman 2002) Sect. 6, where the authors sought to minimize FDR + λFNR), asymptotic limits were not discussed. We point out that the asymptotic limits for the unequally weighted risks are no different from the equally weighted risk, so long as λ is bounded away from zero and infinity. This is because FDR + λFNR vanishes if and only if both FDR and FNR vanish; conversely, non-vanishing FDR and FNR are equivalent to non-vanishing weighted sums. Therefore, a different phase transition would only arise if we weight the Type I and Type II errors by combining family-wise error metrics with marginal error rates.

The next theorem describes the phase transition in the exact–approximate support recovery problem.

Theorem 3.4 *In the context of Theorem 3.2, the function*

$$f_{\mathrm{EA}}(\beta) = 1 \tag{3.40}$$

characterizes the phase transition of the exact–approximate support recovery problem. Namely, the following two results hold:

(i) *If $\underline{r} > f_{EA}(\beta)$, then the procedures listed in Theorem 3.2 with slowly vanishing nominal FWER levels (as defined in Definition 3.1) achieve asymptotically exact–approximate support recovery in the sense of (2.25).*

(ii) *Conversely, if $\overline{r} < f_{EA}(\beta)$, then for any thresholding procedure $\widehat{S}$, the exact–approximate support recovery fails in the sense of (2.26).*

The phase-transition boundary (3.40) is visualized in Fig. 3.2. The proof of this result uses ideas from the proof of Theorem 3.3 and is substantially shorter.

Proof (*Theorem* 3.4) We first show the sufficient condition. Vanishing FWER is guaranteed by the properties of the procedures, and we only need to show that FNR also goes to zero. Similar to the proof of Theorem 3.3, it suffices to show that

$$\text{NDP} = 1 - \widehat{W}_{\text{signal}}(t_p) \to 0, \tag{3.41}$$

where t_p is the threshold of Bonferroni's procedure.

Since α vanishes slowly (see Definition 3.17), for any $\delta > 0$, we have $p^{-\delta} = o(\alpha)$. Therefore, we have $-\log \alpha \le \delta \log p$ for large p, and

$$1 \le \limsup_{p \to \infty} \frac{2 \log p - 2 \log \alpha}{2 \log p} \le 1 + \delta,$$

for any $\delta > 0$. Therefore, by the expression for normal quantiles, we know that

$$t_p = F^{\leftarrow}(1 - \alpha/p) \sim (2 \log p - 2 \log \alpha)^{1/2} \sim (2 \log p)^{1/2}.$$

Since $\underline{r} > f_{EA}(\beta) = 1$, we can pick q such that $1 < q < \underline{r}$. Let $t^* = \sqrt{2q \log p}$, we know that $t_p < t_p^*$ for large p. Therefore, for large p, we have

$$\widehat{W}_{\text{signal}}(t_p) \ge \widehat{W}_{\text{signal}}(t^*) \ge \overline{F_a}(t^*) + o_{\mathbb{P}}(1),$$

where $\overline{F_a}$ is the survival function of $N(\sqrt{2\underline{r} \log p}, 1)$; the last inequality follows from the stochastic monotonicity of the Gaussian location family (2.46), and Lemma 3.1. Indeed, by our choice of $q < \underline{r}$, we obtain

$$F_a(t^*) = \Phi\left(\sqrt{2(q - \underline{r}) \log p}\right) \to 0,$$

and (3.41) is shown. This completes the proof of the sufficient condition.

The proof of the necessary condition follows similar structure as in the proof of Theorem 3.3, and uses the lower bound

$$\text{FWER}(\mathcal{R}) + \text{FNR}(\mathcal{R}) \geq \mathbb{P}\left[\max_{i \in S^c} x(i) > u\right] \wedge \mathbb{E}\left[\frac{|S \setminus \widehat{S}(u)|}{|S|}\right], \tag{3.42}$$

which holds for any arbitrary thresholding procedure $\mathcal{R}$ and arbitrary real $u \in \mathbb{R}$.

By the assumption that $\bar{r} < f_{EA}(\beta) = 1$, we can pick q such that $\bar{r} < q < 1$ and let $u = t^* = \sqrt{2q \log p}$ in (3.42). By the relative stability of iid Gaussian random variables (2.43), we have

$$\mathbb{P}\left[\frac{\max_{i \in S^c} x(i)}{\sqrt{2 \log p}} > \frac{t^*}{\sqrt{2 \log p}}\right] \to 1. \tag{3.43}$$

Since the first fraction in (3.43) converges to 1, while the second converges to $q < 1$. Therefore, the first term on the right-hand side of (3.42) converges to 1.

On the other hand, by the stochastic monotonicity of Gaussian location family (2.46), the probability of missed detection for each signal is lower bounded by $\mathbb{P}[Z + \mu(i) \leq t^*] \geq F_{\bar{a}}(t^*)$, where Z is a standard Gaussian r.v., and $F_{\bar{a}}$ is the cdf of $N(\sqrt{2\bar{r} \log p}, 1)$. Therefore, $|S \setminus \widehat{S}(t^*)| \overset{d}{\geq} \text{Binom}(s, F_{\bar{a}}(t^*))$, and it suffices to show that $F_{\bar{a}}(t^*)$ converges to 1. Indeed,

$$F_{\bar{a}}(t^*) = \Phi(\sqrt{2(q - \bar{r}) \log p}) \to 1,$$

by our choice of $q > \bar{r}$. Hence, both quantities in the minimum on the right-hand side of (3.42) converge to 1 in the limit, and the necessary condition is shown. □

Remark 3.1 The boundary (3.40) was briefly suggested by Arias-Castro and Chen (2017). Unfortunately, it was falsely claimed that the boundary characterized the phase transition of the *exact* support recovery problem, and the alleged proof was left as an "exercise to the reader". This exercise was completed in Chap. 4, where the correct boundary (7.4) was identified.

Theorem 3.4 here shows that the boundary (3.40) *does* exist, though for the different *exact–approximate* support recovery problem. As we will see in Sect. 7.1, the boundary (3.40) also applies to the exact–approximate support recovery problem in chi-square models (1.3).

3.7 The Approximate–Exact Support Recovery Problem

The last phase transition is in terms of the approximate–exact support recovery risk (2.12).

Theorem 3.5 *In the context of Theorem 3.2, the function*

$$f_{AE}(\beta) = \left(\sqrt{\beta} + \sqrt{1 - \beta}\right)^2 \tag{3.44}$$

characterizes the phase transition of the approximate–exact support recovery problem. Namely, the following two results hold:

(i) *If $\underline{r} > f_{AE}(\beta)$, then the Benjamini–Hochberg procedure with slowly vanishing nominal FDR levels (as defined in Definition 3.1) achieves asymptotically approximate–exact support recovery in the sense of (2.25).*

(ii) *Conversely, if $\bar{r} < f_{AE}(\beta)$, then for any thresholding procedure $\widehat{S}$, the approximate–exact support recovery fails in the sense of (2.26).*

The phase-transition boundary (3.44) is visualized in Fig. 3.2.

Proof (*Theorem* 3.5) We first show the sufficient condition (part (i)). Since FDR control is guaranteed by the BH procedure, we only need to show that the FWNR also vanishes, that is,

$$\mathbb{P}\left[\min_{i \in S} x(i) \geq \tau\right] \to 1, \tag{3.45}$$

where τ is the threshold for the BH procedure.

By the assumption that $\underline{r} > f_{AE}(\beta) = (\sqrt{\beta} + \sqrt{1-\beta})^2$, we have $\sqrt{\underline{r}} - \sqrt{1-\beta} > \sqrt{\beta}$, so we can pick $q > 0$, such that

$$\sqrt{\underline{r}} - \sqrt{1-\beta} > \sqrt{q} > \sqrt{\beta}. \tag{3.46}$$

We only need to show that with a specific choice of $t^* = \sqrt{2q \log p}$ where

$$\sqrt{\underline{r}} - \sqrt{1-\beta} > \sqrt{q} > \sqrt{\beta}, \tag{3.47}$$

we have both

$$\mathbb{P}\left[\tau \leq t^*\right] \to 1, \tag{3.48}$$

and

$$\mathbb{P}\left[\min_{i \in S} x(i) \geq t^*\right] \to 1, \tag{3.49}$$

so that

$$\mathbb{P}\left[\min_{i \in S} x(i) \geq \tau\right] \geq \mathbb{P}\left[\min_{i \in S} x(i) \geq t^*, \; t^* \geq \tau\right] \to 1.$$

Relation (3.48) follows in exactly the same way (3.34) did in Sect. 3.4. Dividing the left-hand side in Relation (3.49) by $\sqrt{2 \log p}$, we have

$$\frac{\min_{i \in S} x(i)}{\sqrt{2 \log p}} = \frac{\min_{i \in S} \mu(i) + \epsilon(i)}{\sqrt{2 \log p}} \overset{d}{\geq} \frac{\sqrt{2\underline{r} \log p} + \min_{i \in S} \epsilon(i)}{\sqrt{2 \log p}}$$

$$\to -\sqrt{1-\beta} + \sqrt{\underline{r}},$$

where the last convergence follows from the relative stability of iid Gaussians minima (2.44). On the other hand, $t^*/\sqrt{2\log p} = \sqrt{q} < \sqrt{r} - \sqrt{1-\beta}$ by our choice of q, and Relation (3.49) follows.

The necessary condition follows from the lower bound

$$\text{FDR}(\mathcal{R}) + \text{FWNR}(\mathcal{R}) \geq \mathbb{E}\left[\frac{|\widehat{S}(u) \setminus S|}{|\widehat{S}(u) \setminus S| + |S|}\right] \wedge \mathbb{P}\left[\min_{i \in S} x(i) < u\right], \qquad (3.50)$$

which holds for any thresholding procedure $\mathcal{R}$ and for arbitrary $u \in \mathbb{R}$. In particular, we show that both terms in the minimum in (3.50) converge to 1 when we set $u = t^* = \sqrt{2q \log p}$ where

$$\sqrt{r} - \sqrt{1-\beta} < \sqrt{q} < \sqrt{\beta}. \qquad (3.51)$$

On the one hand, we have

$$\frac{\min_{i \in S} x(i)}{\sqrt{2\log p}} \overset{\mathrm{d}}{\leq} \frac{\min_{i \in S} \epsilon(i) + \sqrt{2r \log p}}{\sqrt{2\log p}} \to \sqrt{r} - \sqrt{1-\beta},$$

by relative stability of iid Gaussians (2.44). On the other hand, $t^*/\sqrt{2\log p} = \sqrt{q} > \sqrt{r} - \sqrt{1-\beta}$ by our choice of q; this shows that the second term on the right-hand side of (3.50) converges to 1.

Observe that $|\widehat{S}(t^*) \setminus S|$ has distribution $\text{Binom}(p - s, \overline{\Phi}(t^*))$, and define $X = X_p := |\widehat{S}(t^*) \setminus S|/|S|$, we obtain

$$\mu := \mathbb{E}[X] = (p^\beta - 1)\overline{\Phi}(t^*) \sim (p^\beta - 1)\frac{\phi(t^*)}{t^*}$$

$$\sim \frac{1}{\sqrt{2\pi}}(2q \log p)^{-1/2} p^{\beta - q} \to \infty,$$

where the divergence follows from our choice of $q < \beta$. Using again Relations (3.24) and (3.25), we conclude that the first term on the right-hand side of (3.50) also converges to 1. This completes the proof of the necessary condition. □

3.8   Asymptotic Power Analysis: A Discussion

Theorems 3.2 through 3.5 allow us to asymptotically quantify the required signal sizes in support recovery problems, as well as in the global hypothesis testing problem in the Gaussian additive error model (3.1). Specifically, these results indicate that at all sparsity levels $\beta \in (0, 1)$, the difficulties of the problems in terms of the required signal sizes have the following ordering:

$$f_D(\beta) < f_A(\beta) < f_{EA}(\beta) < f_{AE}(\beta) < f_E(\beta),$$

as previewed in Fig. 3.2. The ordering aligns with our intuition that the required signal sizes must increase as we move from detection to support recovery problems. Similarly, more stringent criteria for error control (e.g., FWER compared to FDR) require larger signals. We can now also compare $f_{EA}(\beta)$ and $f_{AE}(\beta)$, whose ordering may not be clear from this line of reasoning.

Our last comment is on the gap between FDR and FWER under sparsity assumptions. Although it is believed that FWER control is sometimes too stringent compared to, say, FDR control in support recovery problems, the fact that all five thresholds involve the same scaling indicates that the difficulties of the problems (signal detection, and the four support recovery problems) are comparable when signals are very sparse, i.e., when β is close to 1. This is illustrated with the next example.

Example 3.1 (*Power analysis for variable selection*) For Gaussian errors (AGG with $\nu = 2$), when $\beta = 3/4$, the signal detection boundary (3.4) says that signals will have to be at least of magnitude $\sqrt{(\log p)/2}$, while approximate support recovery (3.19) requires signal sizes of at least $\sqrt{3(\log p)/2}$, and exact support recovery (3.18) calls for signal sizes of at least $\sqrt{9(\log p)/2}$. The required signal sizes increase, but are within the same order of magnitude.

If m independent copies $x_1, \ldots, x_m$ of the observations were made on the same set of p locations, then by taking location-wise averages, $\bar{x}_m(j) = \frac{1}{m} \sum_{i=1}^{m} x_i(j)$, we can reduce error standard deviation, and hence boost the signal-to-noise ratio, by a factor of $\sqrt{m}$. By the simple calculations above, if m samples are needed to detect (sparse) signals of a certain magnitude, then $3m$ samples will enable approximate support recovery with false discovery and non-discovery control, and, in fact, $9m$ samples would enable exact support recovery with family-wise error rates control.

On the other hand, the gap between FDR and FWER is much larger when signals are dense. For example, if the signals are only *approximately* sparse, i.e., having a few components above (3.18) but many smaller components above (3.19), then FDR-controlling procedures will discover substantially larger proportion of signals than FWER-controlling procedures.

Indeed, as $\beta \to 0$, the required signal size for approximate support recovery (3.19) tends to 0, while the required signal size for exact support recovery (3.18) tends to 4 in the Gaussian error models. While Example 3.1 indicates that the exact support recovery is not much more stringent than approximate support recovery when signals are sparse, the gap between required signal sizes widens when signals are dense.

Chapter 4
Exact Support Recovery Under Dependence

We focus on exact support recovery problems in this chapter. Recall from Lemma 2.1 that in order to study the asymptotic behaviors of risk^E, it is sufficient to establish minimal conditions under which the support sets can be consistently estimated, i.e.,

$$\mathbb{P}[\widehat{S}_p = S_p] \longrightarrow 1 \quad \text{as } p \to \infty, \tag{4.1}$$

where $\widehat{S}_p$ is an estimate of the true support set S_p of a high-dimensional signal vector μ_p.

We will establish minimal conditions such that (4.1) holds, by generalizing the results we obtained in Chap. 3 to additive error models with relaxed distributional and dependence assumptions on the additive error array.

4.1 Generalizations of Distributional and Dependence Assumptions

Consider the additive error model (1.1) with the triangular array of errors,

$$\mathcal{E} = \left\{ (\epsilon_p(i))_{i=1}^{p}, \ p = 1, 2, \dots \right\}, \tag{4.2}$$

where the $\epsilon_p(i)$'s have common cumulative distribution function $F(x) = \mathbb{P}[\epsilon_p(i) \leq x]$. In contrast to the assumptions in Chap. 3, we only require the errors to have common marginal distributions.

Although our method of analysis applies to all light-tailed error distributions with rapidly varying tails (see Definition 2.7), to be concrete and better convey the main

© The Author(s), under exclusive license to Springer Nature Switzerland AG 2021
Z. Gao and S. Stoev, *Concentration of Maxima and Fundamental Limits in High-Dimensional Testing and Inference*,
SpringerBriefs in Probability and Mathematical Statistics,
https://doi.org/10.1007/978-3-030-80964-5_4

ideas, we will focus on the class of AGG(ν) laws (see Definition 2.6). Extensions of the results to other classes of error models are presented in Sect. B.1.

This generalized distributional assumption on the errors call for a suitable generalization of the signal size parametrization in order to analyze the problem as we did in the previous chapter. As before, we assume the signals in model (1.1) to be a sparse vector $\mu_p = \left(\mu_p(i)\right)_{i=1}^{p}$, with support $S_p := \{i : \mu_p(i) \neq 0\}$. The sparsity of μ_p, with a few exceptions which will be explicitly stated, is parametrized in terms of a *fixed* regularly varying sequence $\{s_p^\dagger\}$ as follows:

$$|S_p| = \lfloor s_p^\dagger \rfloor, \quad \text{where} \quad s_p^\dagger := \ell(p)p^{1-\beta}, \tag{4.3}$$

for some fixed slowly varying function ℓ. Recall that a function ℓ is slowly varying if $\ell(\lambda t)/\ell(t) \to 1$, as $t \to \infty$, for all $\lambda > 0$. As before, the exponent

$$0 < \beta \leq 1$$

controls the sparsity.

We assume that the non-zero entries of μ are positive and take values in the interval $[\underline{\Delta}, \overline{\Delta}) \subset (0, \infty)$. That is, $0 < \underline{\Delta} \leq \mu(i) < \overline{\Delta} \leq +\infty$, for all $i \in S_p$. The lower and upper bounds on the signal sizes $\underline{\Delta}$ and $\overline{\Delta}$ are parametrized as

$$\underline{\Delta} = \underline{\Delta}(p) = (\nu \underline{r} \log p)^{1/\nu} \quad \text{and} \quad \overline{\Delta} = \overline{\Delta}(p) = (\nu \overline{r} \log p)^{1/\nu}, \tag{4.4}$$

with parameters $0 < \underline{r} \leq \overline{r} \leq +\infty$.

We now turn to the dependence conditions. Several authors have studied the support recovery problem in terms of the Hamming loss and obtained minimax-optimality results (see, e.g., Ji and Jin 2012; Genovese et al. 2012; Jin et al. 2014; Butucea et al. 2018). In the special case of Gaussian marginals, Butucea et al. (2018) showed that the boundary (3.18) exists in a minimax sense. That is, when the errors are *independent* Gaussians, the Hamming loss cannot be made to vanish if the signal sizes are sufficiently small by any procedure. Conversely, if signal size falls below, the Hamming loss can be made to vanish for some thresholding procedure. However, as pointed out in Sect. 2.4, vanishing Hamming loss is only sufficient, not necessary for support recovery (4.1), and results on the former do not carry over directly to the study of the exact support recovery problem. More importantly, since the Hamming loss decomposes into expectations on individual terms that are not affected by dependence, Hamming loss-minimax studies do not reveal the difference in probability of support recovery between independent and dependent observations. This prevents one from fully exploring the phase-transition phenomena under other dependence conditions. As a result, so far in the literature, the role of dependence in model (1.1) has remained largely unexplored.

We take a different approach in this text. In particular, we study the exact support recovery problem (4.1) directly, and show that for thresholding procedures the phase-transition phenomena exist universally in a large class of dependence structures, and not just in a minimax sense.

In a first step, we show that in the AGG model under *arbitrary* dependence, under the scaling described in (4.3) and (4.4), the function

$$f_E(\beta) = f_{E,v}(\beta) = (1 + (1 - \beta)^{1/v})^v, \quad v > 0 \tag{4.5}$$

demarcates the region of possibility for the exact support recovery problem. That is, if the signal sizes are above the boundary (i.e., $\underline{r} > f_E(\beta)$), then FWER-controlling procedures with appropriately calibrated levels achieve exact support recovery (Theorem 4.1). We refer to (4.5) as the *strong classification boundary*.

Conversely, we show that for a surprisingly large class of dependence structures characterized by the concept of *uniform relative stability* (URS, see Definition 4.1), when the signal size is below the boundary (i.e., $r < f_E(\beta)$), no thresholding procedure can achieve the asymptotically perfect support recovery. In fact,

$$\mathbb{P}\left[\widehat{S}_p = S_p\right] \longrightarrow 0, \quad \text{as } p \to \infty, \tag{4.6}$$

for all thresholding procedures (Theorem 4.2). These two results show that the thresholding procedures obey a phase-transition phenomenon in a strong, *point-wise* sense over the class of URS dependence structures, and over the class of AGG(v), $v > 0$ error distributions.

4.2 Sufficient Conditions for Exact Support Recovery

Following Butucea et al. (2018), we define the parameter space for the signals μ as

$$\Theta_p^+(\beta, \underline{r}) = \{\mu \in \mathbb{R}^p : \text{ there exists a set } S_p \subseteq \{1, \ldots, p\} \text{ such that } |S_p| \leq s_p^\dagger,$$
$$\mu(i) \geq (v\underline{r} \log p)^{1/v} \text{ for all } i \in S_p, \text{ and } \mu(i) = 0 \text{ for all } i \notin S_p\}, \tag{4.7}$$

where $s_p^\dagger$ is as in (4.3). Our first result states that, when $F \in \text{AGG}(v)$ with $v > 0$, regardless of the error-dependence structure, (asymptotic) perfect support recovery is achieved by applying Bonferroni's procedure with appropriately calibrated FWER, as long as the minimum signal size $\underline{r}$ is above the strong classification boundary (4.5).

Theorem 4.1 *Let the errors have common marginal distribution $F \in \text{AGG}(v)$ with $v > 0$. Let $\widehat{S}_p$ be Bonferroni's procedure (2.21) with vanishing FWER $\alpha = \alpha(p) \to 0$, such that $\alpha p^\delta \to \infty$ for every $\delta > 0$. If*

$$\underline{r} > f_E(\beta) = (1 + (1 - \beta)^{1/v})^v, \tag{4.8}$$

then we have

$$\lim_{p\to\infty} \sup_{\mu\in\Theta_p^+(\beta,\underline{r})} \mathbb{P}[\widehat{S}_p \neq S_p] = 0. \tag{4.9}$$

Proof Throughout the proof, the dependence on p will be suppressed to simplify notations when such omissions do not lead to ambiguity.

Under the AGG(ν) model, it is easy to see from Eq. (2.33) that the thresholds in Bonferroni's procedure are

$$t_p = F^{\leftarrow}(1 - \alpha/p) = (\nu \log{(p/\alpha)})^{1/\nu}(1 + o(1)). \tag{4.10}$$

It is known that Bonferroni's procedure $\widehat{S}_p = \{i : x(i) > t_p\}$ controls the FWER. Indeed,

$$\mathbb{P}[\widehat{S} \subseteq S] = 1 - \mathbb{P}\left[\max_{i\in S^c} x(i) > t_p\right] = 1 - \mathbb{P}\left[\max_{i\in S^c} \epsilon(i) > t_p\right]$$

$$\geq 1 - \sum_{i=1}^{p} \mathbb{P}[\epsilon(i) > t_p] \geq 1 - \alpha(p) \to 1, \tag{4.11}$$

where we used the union bound in the first inequality. Notice that the lower bound (4.11) is independent of the parameter μ (as well as the dependence structures), and hence holds uniformly over the parameter space, i.e.,

$$\lim_{p\to\infty} \inf_{\mu\in\Theta_p^+(\beta,\underline{r})} P[\widehat{S}_p \subseteq S_p] = 1. \tag{4.12}$$

On the other hand, for the probability of no missed detection, we have

$$\mathbb{P}[\widehat{S} \supseteq S] = \mathbb{P}\left[\min_{i\in S} x(i) > t_p\right] = \mathbb{P}\left[\min_{i\in S} x(i) - (\nu\underline{r}\log p)^{1/\nu} > t_p - (\nu\underline{r}\log p)^{1/\nu}\right].$$

Since the signal sizes are no smaller than $(\nu\underline{r}\log p)^{1/\nu}$, we have

$$x(i) - (\nu\underline{r}\log p)^{1/\nu} \geq \epsilon(i), \quad \text{for all } i \in S,$$

and hence we obtain

$$\mathbb{P}[\widehat{S} \supseteq S] \geq \mathbb{P}\left[\min_{i\in S} \epsilon(i) > (\nu \log{(p/\alpha)})^{1/\nu}(1 + o(1)) - (\nu\underline{r}\log p)^{1/\nu}\right], \tag{4.13}$$

where we plugged in the expression for t_p in (4.10). Now, since the minimum signal size is bounded below by $\underline{r} > (1 + (1 - \beta)^{1/\nu})^{\nu}$, we have $\underline{r}^{1/\nu} - (1 - \beta)^{1/\nu} > 1$, and so we can pick a $\delta > 0$ such that

$$\delta < (\underline{r}^{1/\nu} - (1 - \beta)^{1/\nu})^{\nu} - 1. \tag{4.14}$$

Since by assumption, for all $\delta > 0$, we have $p^{-\delta} = o(\alpha(p))$, there is an $M = M(\delta)$ such that $p/\alpha(p) < p^{1+\delta}$ for all $p \geq M$. Thus, from (4.13), we further conclude that for $p \geq M$ we have

$$\mathbb{P}\left[\widehat{S} \supseteq S\right] \geq \mathbb{P}\left[\min_{i \in S} \epsilon(i) > ((1+\delta)v \log p)^{1/v}(1 + o(1)) - (v\underline{r} \log p)^{1/v}\right]$$

$$= \mathbb{P}\left[\max_{i \in S}(-\epsilon(i)) < \underbrace{(\underline{r}^{1/v} - (1+\delta)^{1/v})(v \log p)^{1/v}(1 + o(1))}_{=:A}\right]$$

$$\geq 1 - \ell(p)p^{1-\beta} \times \overline{F}_-(A), \tag{4.15}$$

where $\overline{F}_-(x) = \mathbb{P}[-\epsilon(i) > x]$ is the survival function of the $(-\epsilon(i))$'s. Notice that (4.15) follows from the union bound and the assumption that $|S_p| \leq s_p^\dagger = \ell(p)p^{1-\beta}$ in (4.7). Therefore, the lower bound does not depend on μ (nor on the error-dependence structure), and holds uniformly in the parameter space. In turn, we obtain

$$\inf_{\mu \in \Theta_p^+(\beta,\underline{r})} \mathbb{P}[\widehat{S}_p \supseteq S_p] \geq 1 - \ell(p)p^{1-\beta} \times \overline{F}_-(A). \tag{4.16}$$

We first show that the right-hand side of (4.16) converges to 1 when $\beta = 1$. Indeed, since $F \in \mathrm{AGG}(\mu)$, we have, for sufficiently large p,

$$\overline{F}_-(A) \leq \overline{F}_-(c(v \log(p))^{1/v}) = O(p^{-c'}),$$

for some $c > c' > 0$. On the other hand, the celebrated Potter bounds for slowly varying functions (see, e.g., Bingham et al. 1987) entail $\ell(p) = o(p^{c'})$, for every $c' > 0$ and hence $\ell(p)\overline{F}_-(A) \to 0$, as $p \to \infty$.

Let now $\beta \in (0, 1)$ and $u_p^- := F_-^{\leftarrow}(1 - 1/p)$. The fact that $p\overline{F}_-(u_p^-) \leq 1$ implies

$$s_p^\dagger \times \overline{F}_-(A) \leq \frac{\overline{F}_-\left(B \times u_{s_p^\dagger}^-\right)}{\overline{F}_-\left(u_{s_p^\dagger}^-\right)}, \tag{4.17}$$

where $B := A/u_{s_p^\dagger}^-$.

Notice that by assumption, the $-\epsilon(i)$'s are also $\mathrm{AGG}(v)$ distributed, and by Proposition 2.1, $u_p^- := F_-^{\leftarrow}(1 - 1/p) \sim (v \log(p))^{1/v}$, as $p \to \infty$. Therefore, we have

$$u_{s_p^\dagger}^- \equiv u_{\ell(p)p^{1-\beta}}^- \sim (v(1-\beta) \log p)^{1/v}, \tag{4.18}$$

where we used the fact that $\log(\ell(p)) = o(\log(p))$. Hence,

$$B = \frac{A}{u_{s_p^\dagger}^-} = \frac{\underline{r}^{1/v} - (1+\delta)^{1/v}}{(1-\beta)^{1/v}}(1 + o(1)) \to c > 1$$

as $p \to \infty$, by our choice of δ in (4.14).

Finally, since the distribution F_- has *rapidly varying* tails (by Definition 2.7 and Example 2.1), applying Proposition 2.2, we conclude that (4.17) vanishes. Consequently, the lower bound on the right-hand side of (4.16) converges to 1. This, combined with (4.12), entails $\lim_{p\to\infty} \inf_{\mu \in \Theta_p^+(\beta, \underline{r})} \mathbb{P}[\widehat{S}_p = S_p] = 1$, and hence the desired Conclusion (4.9), which completes the proof.   □

We end this section with several comments and applications of Theorem 4.1.

Corollary 4.1 (Classes of procedures attaining the boundary) *Relation (4.9) holds for any FWER-controlling procedure that is strictly more powerful than Bonferroni's procedure. This includes Holm's procedure (Holm 1979), and in the case of independent errors, Hochberg's procedure (Hochberg 1988), and the Šidák procedure (Šidák 1967).*

Example 4.1 Under Gaussian errors, the particular choice of the thresholding at $t_p = \sqrt{2 \log p}$ in (2.21) corresponds to a Bonferroni's procedure with FWER decreasing at a rate of $O((\log p)^{-1/2})$, and hence Theorem 4.1 applies. By Corollary 4.1, Holm's procedure—and when the errors are independent, the Šidák, and Hochberg procedures—with FWER controlled at $(\log p)^{-1/2}$ all achieve perfect support recovery provided that $\underline{r} > f_E(\beta)$.

Proof (*Example 4.1*) By Mill's ratio for the standard Gaussian distribution,

$$\frac{t_p \mathbb{P}[Z > t_p]}{\phi(t_p)} \to 1, \quad \text{as} \quad t_p \to \infty,$$

where $Z \sim N(0, 1)$. Using the expression for $t_p = \sqrt{2 \log p}$, we have

$$p\, \mathbb{P}[Z > t_p] \sim \sqrt{2\pi}^{-1} (2 \log p)^{-1/2} \to 0,$$

as desired. The rest of the claims follow from Corollary 4.1.   □

The statements in Theorem 4.1 can be strengthened, to prepare us for a minimax result given in Sect. 5.5.

Remark 4.1 In the proof of Theorem 4.1, both (4.11) and (4.15) hold uniformly over all error-dependence structures. Therefore, (4.12) and (4.16) may be strengthened to yield

$$\lim_{p\to\infty} \sup_{\substack{\mu \in \Theta_p^+(\beta, \underline{r}) \\ \mathcal{E} \in D(F)}} P[\widehat{S}_p \neq S_p] = 0, \tag{4.19}$$

for $\underline{r} > f_E(\beta)$, where $D(F)$ is the collection of all arrays with common marginal F, i.e.,

$$D(F) = \{\mathcal{E} = (\epsilon_p(i))_p : \epsilon_p(i) \sim F \text{ for all } i = 1, \dots, p, \text{ and } p = 1, 2, \dots\}. \tag{4.20}$$

Remark 4.2 We emphasize that Theorem 4.1 holds for errors with *arbitrary* dependence structures. Intuitively, this is because the maxima of the errors grow at their fastest in the case of independence (recall Remark 2.1). Formally, the light-tailed nature of the error distribution allowed us to obtain sharp tail estimates via simple union bounds, valid under arbitrary dependence.

4.3 Dependence and Uniform Relative Stability

An important ingredient needed for a converse of Theorem 4.1 is an appropriate characterization of the error-dependence structure under which the strong classification boundary (4.5) is tight. The notion of *uniform relative stability* turns out to be the key.

Definition 4.1 (*Uniform Relative Stability*) Under the notations established in Definition 2.8, the triangular array $\mathcal{E}$ is said to have uniform relatively stable (URS) maxima if for *every* sequence of subsets $S_p \subseteq \{1, \ldots, p\}$ such that $|S_p| \to \infty$, we have

$$\frac{1}{u_{|S_p|}} M_{S_p} := \frac{1}{u_{|S_p|}} \max_{i \in S_p} \epsilon_p(i) \xrightarrow{\mathbb{P}} 1, \qquad (4.21)$$

as $p \to \infty$, where u_q, $q \in \{1, \ldots, p\}$ is the generalized quantile in (2.37). The collection of arrays $\mathcal{E} = \{\epsilon_p(i)\}$ with URS maxima is denoted $U(F)$.

Uniform relative stability is, as its name suggests, a stronger requirement on dependence than relative stability (recall Definition 2.8). Proposition 2.2 states that an array with iid components sharing a marginal distribution F with rapidly varying tails (Definition 2.7) has relatively stable maxima; it is easy to see that URS also follows, by independence of the entries.

Corollary 4.2 *An independent array $\mathcal{E}$ with common marginals $F \in AGG(v)$, $v > 0$, is URS; in this case, URS holds with $u_{|S_p|} \sim \left(v \log |S_p| \right)^{1/v}$.*

On the other hand, RS and URS hold under much broader dependence structures than just independent errors. These conditions are extremely mild and can be shown to hold for many classes of error models. In Chap. 6, we will focus extensively on the Gaussian case, which is of great interest in applications and is rather challenging. We will provide a simple necessary and sufficient condition for uniform relative stability in terms of the covariance structures.

The relative stability concepts are important because they characterize the dependence structures under which the maxima of error sequences *concentrate* around the quantiles (2.37) in the sense of (2.38). This concentration of maxima phenomena, in turn, is the key to establishing the necessary conditions of the phase-transition results in support recovery problems.

4.4 Necessary Conditions for Exact Support Recovery

With the preparations from Sect. 4.3, we are ready to state the necessary conditions for exact support recovery (4.1) by thresholding procedures. It turns out that the strong classification boundary (4.5) is tight, under the general dependence structure characterized by URS (Definition 4.1).

Formally, we define the parameter space for the signals μ to be

$$\Theta_p^-(\beta, \bar{r}) = \{\mu \in \mathbb{R}^p : \text{ there exists a set } S_p \subseteq \{1, \ldots, p\} \text{ such that } |S_p| = \lfloor s_p^\dagger \rfloor,$$
$$0 < \mu(i) \leq (\nu \bar{r} \log p)^{1/\nu} \text{ for all } i \in S_p, \text{ and } \mu(i) = 0 \text{ for all } i \notin S_p\}, \tag{4.22}$$

where $s_p^\dagger = \ell(p) p^{1-\beta}$ is as in (4.3).

Theorem 4.2 *Let $\mathcal{E}$ be a triangular array with common AGG(ν) marginal F, $\nu > 0$. Assume further that the errors $\mathcal{E}$ have uniform relatively stable maxima and minima, i.e., $\mathcal{E} \in U(F)$, and $(-\mathcal{E}) = \{-\epsilon_p(i)\} \in U(F)$. If*

$$\bar{r} < f_E(\beta) = \left(1 + (1-\beta)^{1/\nu}\right)^\nu, \tag{4.23}$$

then

$$\lim_{p \to \infty} \inf_{\widehat{S}_p \in \mathcal{T}} \inf_{\mu \in \Theta_p^-(\beta, \bar{r})} \mathbb{P}[\widehat{S}_p \neq S_p] = 1, \tag{4.24}$$

where $\mathcal{T}$ is the class of all thresholding procedures (2.20).

Proof To avoid cumbersome double subscript notations, we will sometimes suppress dependence on p of the set sequences $\widehat{S}_p$ and S_p in the proof.

Since the estimator $\widehat{S}_p = \{x(i) \geq t_p(x)\}$ is thresholding, exact support recovery takes place if and only if the threshold separates the signals and the null part, i.e.,

$$\mathbb{P}[\widehat{S}_p = S_p] = \mathbb{P}\left[\max_{i \in S^c} x(i) < t_p(x) \leq \min_{i \in S} x(i)\right] \leq \mathbb{P}\left[\max_{i \in S^c} x(i) < \min_{i \in S} x(i)\right].$$

Since the right-hand side does not depend on the procedure $\widehat{S}_p$, we also have

$$\sup_{\widehat{S}_p \in \mathcal{T}} \mathbb{P}[\widehat{S}_p = S_p] \leq \mathbb{P}\left[\max_{i \in S^c} x(i) < \min_{i \in S} x(i)\right] \leq \mathbb{P}\left[\max_{i \in S^c} \epsilon(i) < \overline{\Delta} + \min_{i \in S} \epsilon(i)\right], \tag{4.25}$$

where we used the assumption that the signal sizes are no greater than $\overline{\Delta}$. Let $S^* = S_p^*$ be a sequence of support sets that maximize the right-hand side of (4.25), i.e., let

$$S_p^* = \underset{S \subseteq \{1, \ldots, p\}: |S| = \lfloor s_p^\dagger \rfloor}{\arg \max} \mathbb{P}\left[\max_{i \in S^c} \epsilon(i) < \overline{\Delta} + \min_{i \in S} \epsilon(i)\right],$$

where $s_p^\dagger = \ell(p)p^{1-\beta}$ is the size of the true support set, and ties are broken lexico-graphically if multiple maximizers exist. Then, we obtain the following bound which only depends on $\bar{r}$ and the distribution of $\mathcal{E}$,

$$\sup_{\widehat{S}_p \in \mathcal{T}} \sup_{\mu \in \Theta_p^-(\beta, \bar{r})} \mathbb{P}[\widehat{S}_p = S_p] \le \mathbb{P}\left[\max_{i \in S^{*c}} \epsilon(i) < \overline{\Delta} + \min_{i \in S^*} \epsilon(i)\right]$$

$$= \mathbb{P}\left[\frac{M_{S^{*c}}}{u_p} < \frac{\overline{\Delta} - m_{S^*}}{u_p}\right], \qquad (4.26)$$

where $M_{S^{*c}} = \max_{i \in S^{*c}} \epsilon(i)$ and $m_{S^*} = \max_{i \in S^*}(-\epsilon(i))$. Since the error arrays $\mathcal{E}$ and $(-\mathcal{E})$ are URS by assumption, using the expression for the AGG quantiles (2.33), we have

$$\frac{M_{S^{*c}}}{u_p} = \frac{M_{S^{*c}}}{u_{|S^{*c}|}} \frac{u_{|S^{*c}|}}{u_p} \xrightarrow{\mathbb{P}} 1, \quad \text{and} \quad \frac{m_{S^*}}{u_p} = \frac{m_{S^*}}{u_{|S^*|}} \frac{u_{|S^*|}}{u_p} \xrightarrow{\mathbb{P}} (1-\beta)^{1/\nu}, \qquad (4.27)$$

so that the two random terms in probability (4.26) converge to constants. Notice that the second relation in (4.27) holds by URS for any $\beta \in (0, 1)$. When $\beta = 1$, the relation holds because $\{m_{S^*}/u_{|S^*|}\}$ is tight, while $0 \le u_{|S^*|}/u_p \le u_{\ell(p)}/u_p \to 0$ since $\ell(p) = o(p)$ by the Potter bounds for slowly varying functions (see, e.g., Bingham et al. 1987).

Since signal sizes are bounded above by $\bar{r} < \left(1 + (1-\beta)^{1/\nu}\right)^\nu$, we can write $\bar{r}^{1/\nu} = 1 + (1-\beta)^{1/\nu} - d$ for some $d > 0$. By our parametrization of $\overline{\Delta}$, we have

$$\frac{\overline{\Delta}}{u_p} = \left(1 + (1-\beta)^{1/\nu} - d\right)(1 + o(1)). \qquad (4.28)$$

Combining (4.27) and (4.28), we conclude that the right-hand side of the probability (4.26) converges in probability to a constant strictly less than 1, that is,

$$\frac{\overline{\Delta} - m_S}{u_p} \xrightarrow{\mathbb{P}} 1 - d, \qquad (4.29)$$

while $M_{S^{*c}}/u_p \xrightarrow{\mathbb{P}} 1$. Therefore, the probability in (4.26) must go to 0. $\qquad \square$

We end this section with several remarks on the scope and consequences of our results. Our first comment is on the signal sizes and, in particular, on the gap between the sufficient conditions (Theorem 4.1) and the necessary conditions (Theorem 4.2).

Remark 4.3 (*Minding the gap*) The sufficient condition in Theorem 4.1 requires that *all* signals be larger than the strong classification boundary $f_E(\beta)$ in order to achieve exact support recovery (4.1), while Theorem 4.2 states that exact support recovery fails (in the sense of (4.6)) when *all* signal sizes are below the boundary— the two conditions are *not* complements of each other. This gap between the sufficient and necessary conditions on signal sizes, however, may be difficult to bridge. Indeed, in general, when signal sizes straddle the boundary $f_E(\beta)$, either outcome is possible, as we demonstrate in Example 4.2.

Example 4.2 (*Signals straddling the boundary*) Let the signal μ have $|S_p| = \lfloor p^{(1-\beta)} \rfloor$ non-zero entries composed of two disjoint sets $S_p = S_p^{(1)} \cup S_p^{(2)}$. Let also the magnitude of the signals be equal within the two sets, i.e., $\mu(i) = \sqrt{2r^{(k)} \log p}$ if $i \in S_p^{(k)}$ for some constants $r^{(k)} > 0$ for $k = 1, 2$. For simplicity, assume that the errors are iid standard Gaussians.

Consider two scenarios

1. $r^{(1)} = (1 + \delta) f_E(\beta)$, $r^{(2)} = (1 + \delta)$, with $|S_p^{(1)}| = |S_p| - 1$, $|S_p^{(2)}| = 1$,
2. $r^{(1)} = (1 + \delta) f_E(\beta)$, $r^{(2)} = (1 - \delta) f_E(\beta)$, with $|S_p^{(1)}| = \lfloor |S_p|/2 \rfloor$, $|S_p^{(2)}| = |S_p| - |S_p^{(1)}|$

for some constants $0 < \delta < 1 - \beta < 1$. In both cases, the signals in $S_p^{(1)}$ (respectively, $S_p^{(2)}$) are above (respectively, below) the strong classification boundary (4.5). However, in the first scenario, we have $\mathbb{P}[\widehat{S}_p^{\mathrm{Bonf}} = S_p] \to 1$ where $\widehat{S}_p^{\mathrm{Bonf}}$ is Bonferroni's procedure described in Theorem 4.1, while in the second scenario, we have $\mathbb{P}[\widehat{S}_p = S_p] \to 0$ for *all* thresholding procedures $\widehat{S}_p$.

Proof (*Example 4.2*) In the first scenario, the signal sizes in $S_p^{(1)}$ are by definition above the strong classification boundary (4.5). The signal in $S_p^{(2)}$ has size parameter $1 + \delta < 2 - \beta < (1 + \sqrt{1 - \beta})^2$, and therefore falls below the boundary.

It remains to show that $\mathbb{P}[\widehat{S}_p^{\mathrm{Bonf}} = S_p] \to 1$. To do so, we define two new arrays

$$\mathcal{Y}^{(k)} = \{y_p^{(k)}(j), \ j = 1, 2, \ldots, p\}, \quad k \in \{1, 2\}_p,$$

where $y_p^{(k)}(j) = x_p(j)$ if $j \notin S_p^{(k)}$, and $y_p^{(k)}(j) = \tilde{\epsilon}_p(j)$ if $j \in S_p^{(k)}$, using an independent error array $\{\tilde{\epsilon}_p(j), \ j = 1, \ldots, p\}$ with iid standard Gaussian elements. That is, we replace the elements in $S_p^{(1)}$ and $S_p^{(2)}$ with iid standard Gaussian noise. Notice both arrays $\mathcal{Y}^{(1)}$ and $\mathcal{Y}^{(2)}$ satisfy the conditions in Theorem 4.1 (with sparsity parameter equal to β and 1, respectively). Hence, we have

$$\mathbb{P}[\widehat{S}_p^{\mathrm{Bonf}} \subseteq S_p] = \mathbb{P}\left[\max_{j \in S^c} x(j) \le t_p\right] \le \mathbb{P}\left[\max_{j \in S^c} y^{(1)}(j) \le t_p\right] \to 0,$$

and

$$\mathbb{P}[\widehat{S}_p^{\mathrm{Bonf}} \supseteq S_p] = \mathbb{P}\left[\min_{j \in S} x(j) > t_p\right] \ge 1 - \mathbb{P}\left[\min_{j \in S^{(1)}} x(j) \le t_p\right] - \mathbb{P}\left[\min_{j \in S^{(2)}} x(j) \le t_p\right]$$

$$\ge 1 - \mathbb{P}\left[\min_{j \in S^{(1)}} y_p^{(2)}(j) \le t_p\right] - \mathbb{P}\left[\min_{j \in S^{(2)}} y_p^{(1)}(j) \le t_p\right] \to 1,$$

where t_p is the threshold in Bonferroni's procedure. The conclusion follows.

In the second scenario, the signal sizes in $S^{(2)}$ by definition fall below the strong classification boundary (4.5). To see that no thresholding procedure succeeds, we adapt the proof of Theorem 4.2. In particular, we obtain

$$\mathbb{P}[\widehat{S}_p = S_p] \leq \mathbb{P}\left[\max_{j \in S^c} x(j) \leq t_p < \min_{j \in S} x(j)\right] \leq \mathbb{P}\left[\max_{j \in S^c} x(j) < \min_{j \in S^{(2)}} x(j)\right].$$

By the assumption that signals in $S^{(2)}$ have size parameter $(1-\delta)f_E(\beta)$, we have

$$\mathbb{P}\left[\max_{j \in S^c} x(j) < \min_{j \in S^{(2)}} x(j)\right] = \mathbb{P}\left[\frac{M_{S^c}}{u_p} < \frac{\sqrt{2(1-\delta)f_E(\beta)\log p} - m_{S^{(2)}}}{u_p}\right],$$
(4.30)

where $M_{S^c} = \max_{j \in S^c} \epsilon(j)$ and $m_{S^{(2)}} = \max_{j \in S^{(2)}}(-\epsilon(j))$. The ratio on the left-hand side of the inequality converges to 1 as in (4.27), whereas the term on the right-hand side

$$\frac{\sqrt{2(1-\delta)f_E(\beta)\log p} - m_{S^{(2)}}}{u_p} = \sqrt{(1-\delta)f_E(\beta)} - \frac{m_{S^{(2)}}}{u_{|S^{(2)}|}} \frac{u_{|S^{(2)}|}}{u_p}$$
$$\xrightarrow{\mathbb{P}} \sqrt{(1-\delta)} + \sqrt{1-\beta}(\sqrt{(1-\delta)} - 1) < 1,$$

where we used the URS of the error arrays, and that

$$u_{|S^{(2)}|} \sim \sqrt{2\log(p^{1-\beta}/2)} = \sqrt{2((1-\beta)\log p - \log 2)} \sim \sqrt{2(1-\beta)\log p}$$

to conclude the convergence in probability. □

Our second remark is on the restriction to thresholding procedures.

Remark 4.4 Since the sharp phase-transition result just established applies only to the general class of thresholding procedures, it is natural to ask if other good procedures have been left out by this restriction. We will establish later in Chap. 5 that in many cases the optimal procedures are, in fact, thresholding procedures. In general, however, thresholding procedures can be sub-optimal, e.g., when the errors have heavy (regularly varying) tails. We will also demonstrate the absence of a phase-transition phenomenon in exact support recovery by thresholding, in Supplement Sect. B.2.

Our final comment is on the interplay between thresholding procedures and the dependence class characterized by URS.

Remark 4.5 Paraphrasing Theorems 4.1 and 4.2: if we consider only thresholding procedures, then for a very large class of dependence structures, we cannot improve upon the Bonferroni procedure $\widehat{S}_p^{\text{Bonf}}$. Specifically, for all $\mathcal{E} \in U(F)$ and $-\mathcal{E} \in U(F)$, and for all $S_p \in \mathcal{S}$, where $\mathcal{S} = \{S \subseteq \{1, \ldots, p\}; |S| = \lfloor \ell(p)p^{1-\beta} \rfloor\}$, we have

$$\lim_{p\to\infty} \mathbb{P}[\widehat{S}_p^{\text{Bonf}} \neq S_p] = \begin{cases} \lim\sup_{p\to\infty} \inf_{\widehat{S}_p\in\mathcal{T}} \mathbb{P}[\widehat{S}_p \neq S_p] = 0, & \text{if } \underline{r} > f_{\text{E}}(\beta), \\ \lim\inf_{p\to\infty} \inf_{\widehat{S}_p\in\mathcal{T}} \mathbb{P}[\widehat{S}_p \neq S_p] = 1, & \text{if } \overline{r} < f_{\text{E}}(\beta), \end{cases}$$

$$(4.31)$$

where $\mathcal{T}$ is the set of all thresholding procedures (2.20).

Theorem 4.2 answers a question raised in Butucea et al. (2018). In particular, the authors of (Butucea et al. 2018) commented that independent error is the "least favorable model" in the problem of support recovery, and conjectured that the support recovery problem may be easier to solve under dependence, similar to how the problem of signal detection is easier under dependent errors (Hall and Jin 2010). Surprisingly, our results here state that asymptotically *all* error-dependence structures in the large URS class are equally difficult for thresholding procedures. Therefore, the phase-transition behavior is universal in the class of dependence structures characterized by URS.

We emphasize the restriction to the URS dependence class in Theorem 4.2 is not an assumption of convenience. The dependence condition characterized by uniform relative stability is, in fact, one of the weakest in the literature. We will partially characterize the URS dependence class in Chap. 6.

4.5 Dense Signals

We treat briefly the case of dense signals, where the size of the support set is proportional to the problem dimension, i.e., $s \sim cp$ for some constant $c \in (0, 1)$. We show that in this case, a phase-transition-type result still holds, independently of the value of c. Analogous to the setup of Theorems 4.1 and 4.2, let

$$\Theta_p^{\text{d}+}(c, \underline{r}) = \{\mu \in \mathbb{R}^p : \text{there exists a set } S_p \subseteq \{1, \dots, p\} \text{ such that } |S_p| \leq \lfloor cp \rfloor,$$
$$\mu(i) \geq (v\underline{r}\log p)^{1/v} \text{ for all } i \in S_p, \text{ and } \mu(i) = 0 \text{ for all } i \notin S_p\},$$

$$(4.32)$$

where "d" in the notation $\Theta_p^{\text{d}+}$ stands for "dense". Similarly, define

$$\Theta_p^{\text{d}-}(c, \overline{r}) = \{\mu \in \mathbb{R}^p : \text{there exists a set } S_p \subseteq \{1, \dots, p\} \text{ such that } |S_p| = \lfloor cp \rfloor,$$
$$0 < \mu(i) \leq (v\overline{r}\log p)^{1/v} \text{ for all } i \in S_p, \text{ and } \mu(i) = 0 \text{ for all } i \notin S_p\}.$$

$$(4.33)$$

Theorem 4.3 *Let $c \in (0, 1)$ be a fixed constant, and let $\widehat{S}_p^{\text{Bonf}}$ denote Bonferroni's procedure as described in Theorem 4.1. In the context of Theorem 4.1, if $\underline{r} > 1$, then we have*

$$\lim_{p\to\infty} \sup_{\mu\in\Theta_p^{\text{d}+}(c,\underline{r})} \mathbb{P}[\widehat{S}_p^{\text{Bonf}} \neq S_p] = 0.$$

$$(4.34)$$

While in the context of Theorem 4.2, if $\bar{r} < 1$, then

$$\lim_{p \to \infty} \inf_{\widehat{S}_p \in \mathcal{T}} \inf_{\mu \in \Theta_p^{d-}(c, \bar{r})} \mathbb{P}[\widehat{S}_p \neq S_p] = 1, \tag{4.35}$$

where $\mathcal{T}$ is the class of all thresholding procedures (2.20).

Remark 4.6 Notice that the boundary for the signal size parameter is identically 1 in this dense regime. Therefore, if we interpret $\beta = 0$ of the parametrization (4.3) as $s \sim cp$, where $c \in (0, 1)$, then the strong classification boundary (4.5) may be continuously extended to the left-end point where $f_E(0) = 1$.

Proof *(Theorem 4.3)* The proof is entirely analogous to that of Theorems 4.1 and 4.2. Specifically, (4.34) follows by replacing $\lfloor \ell(p) p^{1-\beta} \rfloor$ with $\lfloor cp \rfloor$ in Relation (4.15) onward, and replacing (4.18) with

$$u_s^- \sim (v \log cp)^{1/v} \sim (v \log p)^{1/v}$$

in the proof of Theorem 4.1. Similarly, (4.35) follows the proof of Theorem 4.2. Indeed, by using the fact that

$$\frac{u_{|S^{*c}|}}{u_p} \sim \frac{(v \log (1-c)p)^{1/v}}{(v \log p)^{1/v}} \to 1$$

and $u_{|S^*|}/u_p \to 1$ for all $c \in (0, 1)$, we see that Relation (4.27) holds with $\beta = 0$, and the rest of Theorem 4.2 applies. $\qquad\square$

4.6 Numerical Illustrations for Independent Errors

We examine numerically Boundaries (4.5) under several error tail assumptions for independence errors in this section. Numerical experiments for dependent errors will be deferred until we characterize the URS conditions in Chap. 6.

To demonstrate the phase-transition phenomenon under different error tail densities, we simulate from the additive error model (1.1) with

- Gaussian errors, where the density is given by $f(x) = \frac{1}{\sqrt{2\pi}} \exp\{-x^2/2\}$.
- Laplace errors, where the density is given by $f(x) = \frac{1}{2} \exp\{-|x|\}$.
- Generalized Gaussian $v = 1/2$, with density $f(x) = \frac{1}{2} \exp\{-2|x|^{1/2}\}$.

The sparsity and signal size of the sparse mean vector are parametrized as in Eqs. (4.3) and (4.4), respectively. The support set S is estimated with $\widetilde{S} = \{i : x(i) > \sqrt{2 \log p}\}$ under the Gaussian errors, $\widetilde{S} = \{i : x(i) > \log p + (\log \log p)/2\}$ under the Laplace errors, and with $\widetilde{S} = \{i : x(i) > \frac{1}{4} (W(-c/(ep \log p)) + 1)^2\}$ under the generalized Gaussian ($v = 1/2$) errors. Here W is the Lambert W function, i.e., $W = f^{-1}$ where

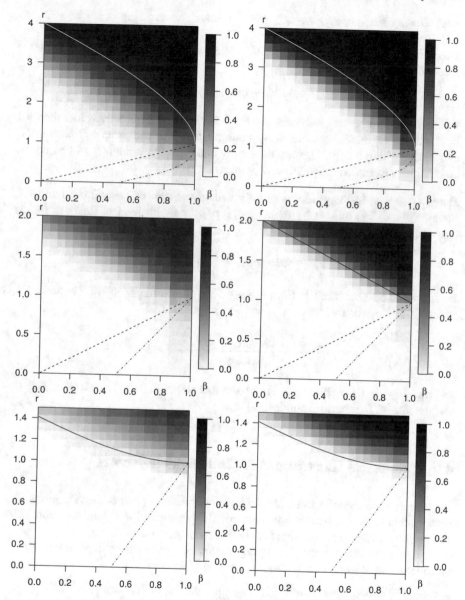

Fig. 4.1 The empirical probability of exact support recovery from numerical experiments, as a function of sparsity level β and signal sizes r, from Gaussian error models (upper panels), Laplace error models (middle panels), and generalized Gaussian with $\nu = 1/2$ (lower panels); darker color indicates higher probability of exact support recovery. The experiments were repeated 1000 times for each sparsity–signal size combination, and for dimensions $p = 100$ (left panels) and $p = 10000$ (right panels). Numerical results agree with the boundaries described in Theorem 4.1; convergence is noticeably slower for under generalized Gaussian ($\nu = 1/2$) errors. For reference, the dashed and dash-dotted lines represent the weak classification and detection boundaries (see Chap. 3)

$f(x) = x \exp(x)$. The choices of thresholds correspond to Bonferroni's procedures with FWER decreasing at a rate of $1/\sqrt{\log p}$, therefore satisfying the assumptions in Theorem 4.1. Experiments were repeated 1000 times under each sparsity and signal size combination.

The results of the numerical experiments are shown in Fig. 4.1. The numerical results illustrate that the predicted boundaries are not only accurate in high dimensions ($p = 10000$, right panels of Fig. 4.1), but also practically meaningful even at moderate dimensions ($p = 100$, left panels of Fig. 4.1).

Chapter 5
Bayes and Minimax Optimality

In this chapter, we investigate the universality of the phase-transition results on exact support recovery established in Chap. 4. Specifically, we would like to know to what extent the strong classification boundary applies to all support estimators and not just thresholding ones. The answer to this question will complete the characterization of the fundamental limits in exact support recovery.

To this end, we begin by characterizing the finite-sample Bayes optimality of the thresholding procedures. As we will see, the so-called oracle thresholding estimators are in fact finite-sample optimal for many classes of models. These optimality results allow us to establish a minimax formulation of the exact support recovery phase-transition phenomenon that covers arbitrary procedures.

Perhaps surprisingly, thresholding estimators can be sub-optimal. This is so, for example, in the additive noise model when the error tail densities are heavier than exponential. In this case, we will see that *likelihood ratio thresholding* rather than *data thresholding* are the optimal support estimators.

5.1 Bayes Optimality in Support Recovery Problems

In studying support recovery problems, restrictions to the thresholding procedures are sometimes justified by arguing that such procedures are the "reasonable" choice for estimating the support set (see, e.g., Arias-Castro and Chen 2017). We show in this chapter that, for general error models, thresholding procedures are not always optimal, even when the observations are independent.

We shall identify the optimal procedure for support recovery problems under a Bayesian setting with general distributional assumptions (including but not limited to additive models (1.1)). Specifically, we assume that there is an ordered set $P = (i_1, \ldots, i_s)$, $i_i \in \{1, \ldots, p\}$, and s not necessarily equal densities $f_1, \ldots, f_s$, such that the observations indexed by set P have corresponding densities. That is,

© The Author(s), under exclusive license to Springer Nature Switzerland AG 2021
Z. Gao and S. Stoev, *Concentration of Maxima and Fundamental Limits in High-Dimensional Testing and Inference*,
SpringerBriefs in Probability and Mathematical Statistics,
https://doi.org/10.1007/978-3-030-80964-5_5

$$x(i_j) \sim f_j, \quad j = 1, \ldots, s. \tag{5.1}$$

Let also the rest $(p - s)$ observations have common density f_0, i.e., $x(i) \sim f_0$ for $i \notin S$. We further assume that the observations x are mutually independent.

We adopt here a Bayesian framework to measure statistical risks. Let the ordered support $P = (i_1, \ldots, i_s)$ have prior

$$\pi((i_1, \ldots, i_s)) = (p - s)!/p!, \tag{5.2}$$

for all distinct $1 \leq i_1 < \ldots < i_s \leq p$. Consequently, the unordered support $S = \{i_1, \ldots, i_s\}$ is distributed uniformly in the collection of all set of size s, with the unordered uniform distribution π^u. That is, for all $S \in \mathcal{S} := \{S \subseteq \{1, \ldots, p\}; |S| = s\}$, we have

$$\pi^u(\{i_1, \ldots, i_s\}) = \sum_{\sigma} \pi((i_{\sigma(1)}, \ldots, i_{\sigma(s)})) = (p - s)!s!/p!, \tag{5.3}$$

where the sum is taken over all permutations of $\{1, 2, \ldots, s\}$.

For any fixed configuration P, consider the loss function,

$$\ell(\widehat{S}, S) := \mathbb{P}[\widehat{S} \neq S] = \mathbb{P}_P[\widehat{S} \neq S],$$

where the probability is taken over the randomness in the observations x only. The Bayes optimal procedures, by definitions, should minimize

$$\mathbb{E}_{\pi}\mathbb{P}[\widehat{S} \neq S], \tag{5.4}$$

where the expectation is taken over the random configurations P, with a uniform distribution π as specified in (5.2).

If, however, the sparsity $s = |S|$ of the problem is known, then a "natural" estimator for S would be based on the set of top s-order statistics. Such estimators will be referred to as oracle thresholding estimators and formally defined next.

For any collection of numbers $\{a_i, i = 1, \cdots, s\}$, let

$$\langle a_1, \cdots, a_s \rangle := (a_{[1]}, \cdots, a_{[s]})$$

denote the vector of a_i's arranged in a non-increasing order.

Definition 5.1 (*Oracle data thresholding*) Let $x_{[1]} \geq \ldots \geq x_{[p]}$ be the order statistics of the data vector x. Any estimator $\widehat{S}^* := \{i_1, \cdots, i_s\}$, where

$$\langle x(i_1), \cdots, x(i_s) \rangle = (x_{[1]}, \cdots, x_{[s]})$$

will be referred to as an *oracle thresholding estimator*.

Simply put, the oracle thresholding estimators are comprised of the indices corresponding to the s largest values in the data. Note that, in the absence of ties among the largest $s + 1$ data values, the oracle thresholding estimator is unique. For concreteness, one can break possible ties lexicographically. In many cases, the oracle thresholding estimators will be almost surely unique.

5.2 Bayes Optimality of Oracle Thresholding

In this section, we study the Bayes optimality of the oracle thresholding procedures. The following *monotone likelihood ratio* (MLR) property will play a key role.

Definition 5.2 *(Monotone Likelihood Ratio)* A family of positive densities on $\mathbb{R}$, $\{f_\delta, \delta \in U\}$, is said to have the MLR property if, for all $\delta_0, \delta_1 \in U \subseteq \mathbb{R}$ such that $\delta_0 < \delta_1$, the likelihood ratio $\big(f_{\delta_1}(x)/f_{\delta_0}(x)\big)$ is an increasing function of x.

The next result provides a general criterion for the finite-sample Bayes optimality of the oracle thresholding procedure $\widehat{S}^*$.

Theorem 5.1 *Let the observations $x(i)$, $i = 1, \ldots, p$ be as prescribed as in (5.1) through (5.2). If each of the pairs $\{f_0, f_1\}, \ldots, \{f_0, f_s\}$ forms an MLR family, then every oracle data thresholding procedure $\widehat{S}^*$ is finite-sample optimal in terms of Bayes risk $\mathbb{E}_\pi \mathbb{P}[\widehat{S} \neq S]$. That is,*

$$\widehat{S}^* \in \arg\min_{\widehat{S}} \mathbb{E}_\pi \mathbb{P}[\widehat{S} \neq S]. \tag{5.5}$$

for all s and p.

Proof The problem of support recovery can be equivalently stated as a classification problem, where the discrete parameter space is $\mathcal{S} = \{S \subseteq \{1, \ldots, p\} : |S| = s\}$, and the observation $x \in \mathbb{R}^p$ has likelihood $f(x|S)$ indexed by the support set S.

By the optimality of the Bayes classifier (see, e.g., Domingos and Pazzani 1997), a set estimator that maximizes the probability of support recovery is one such that

$$\widehat{S} \in \arg\max_{S \in \mathcal{S}} f(x|S)\pi(S).$$

Since we know from (5.3) that $\pi(\cdot)$ is uniform, the problem in our context reduces to showing that $f(x|\widehat{S}^*) = f(x|\widehat{S})$, where $f(x|S)$ is the conditional distribution of data given the unordered support S,

$$f(x|S) = \sum_{P \in \sigma(S)} f(x|P)\pi^{\text{ord}}(P|S) = \frac{1}{s!}\left(\sum_{P \in \sigma(S)} \prod_{i=1}^{s} f_i(x(P(i)))\right) \prod_{k \notin S} f_0(x(k)),$$

where $\sigma(S)$ is the set of all permutations of the indices in the support set S.

Suppose that $\widehat{S}$ is *not* an oracle thresholding estimator, then there must be indices $j \in \widehat{S}$ and $j' \notin \widehat{S}$ such that $x(j) < x(j')$. We exchange the classifications of $x(j)$ and $x(j')$, and form a new estimate $\widehat{S}' = (\widehat{S} \setminus \{j\}) \cup \{j'\}$. Comparing the likelihoods under $\widehat{S}$ and $\widehat{S}'$, we have

$$
\begin{aligned}
f(x|\widehat{S}) - f(x|\widehat{S}') &= \frac{1}{s!} \sum_{P \in \sigma(\widehat{S})} \prod_{i=1}^{s} f_i(x(P(i))) f_0(x(j')) \prod_{k \notin \widehat{S} \cup \{j'\}} f_0(x(k)) - \\
&\quad - \frac{1}{s!} \sum_{P' \in \sigma(\widehat{S}')} \prod_{i=1}^{s} f_i(x(P'(i))) f_0(x(j)) \prod_{k \notin \widehat{S}' \cup \{j\}} f_0(x(k)) \\
&= \frac{1}{s!} \left(\sum_{i=1}^{s} a_i \Big(f_i(x(j)) f_0(x(j')) - f_i(x(j')) f_0(x(j)) \Big) \right) \prod_{k \notin \widehat{S} \cup \{j'\}} f_0(x(k)),
\end{aligned}
$$
(5.6)

where the last equality follows by first summing over all permutations fixing $P(i) = j$ and $P'(i) = j'$, and setting $a_i = \sum_{P \in \sigma(\widehat{S} \setminus \{j\})} \prod_{i' \neq i} f_{i'}(x(P(i')))$. Notice that the a_i's are non-negative.

Since $x(j) < x(j')$, and since each of $\{f_0, f_i\}$ is an MLR family, we have

$$
\frac{f_i(x(j))}{f_0(x(j))} - \frac{f_i(x(j'))}{f_0(x(j'))} \leq 0 \implies f_i(x(j)) f_0(x(j')) - f_i(x(j')) f_0(x(j)) \leq 0.
$$

Using Relation (5.6), we conclude that $f(x|\widehat{S}) \leq f(x|\widehat{S}')$. Continuing this way, we can successively improve the likelihood of every estimator until we arrive at an oracle thresholding estimator, proving the desired optimality. Note that with the same argument, we obtain that any two oracle thresholding estimators have the same likelihood.   □

We emphasize that under the MLR conditions in Theorem 5.1, the oracle thresholding procedures are in fact *finite-sample optimal* in the above Bayesian context. Further, our setup allows for different alternative distributions, and relaxes the assumptions of Butucea et al. (2018), where the alternatives are assumed to be identically distributed.

It remains to understand when the key MLR property holds. We elaborate on this question next. Returning to the more concrete signal-plus-noise model (1.1), it turns out that the error tail behavior is what determines the optimality of data thresholding procedures. In this setting, log-concavity of the error densities is *equivalent* to the MLR property (Lemma 5.1). This, in turn, yields the finite-sample optimality of data thresholding procedures (Corollary 5.1).

Lemma 5.1 *Let δ be the magnitude of the non-zero signals in the signal-plus-noise model (1.1) with positive error density f_0, and let $f_\delta(x) = f_0(x - \delta)$. The family $\{f_\delta, \delta \in \mathbb{R}\}$ has the MLR property if and only if the error density f_0 is log-concave.*

Proof Suppose MLR holds, we will show that $f_0(t) = \exp\{\phi(t)\}$ for some concave function ϕ. By the assumption of MLR, for any $x_1 < x_2$, setting $\delta_0 = 0$, and $\delta_1 = (x_2 - x_1)/2 > 0$, we have

$$\log \frac{f_{\delta_1}(x_2)}{f_{\delta_0}(x_2)} = \phi\left(\frac{(x_1 + x_2)}{2}\right) - \phi(x_2) \geq \phi(x_1) - \phi\left(\frac{(x_1 + x_2)}{2}\right) = \log \frac{f_{\delta_1}(x_1)}{f_{\delta_0}(x_1)}.$$

This implies that the log-density $\phi(t)$ is midpoint-concave, i.e., for all x_1 and x_2, we have

$$\phi\left(\frac{(x_1 + x_2)}{2}\right) \geq \frac{1}{2}\phi(x_1) + \frac{1}{2}\phi(x_2). \tag{5.7}$$

For Lebesgue measurable functions, midpoint concavity is equivalent to concavity by the Sierpinski Theorem see, (e.g., Sec I.3 of Donoghue 2014). This proves the "only-if" part.

For the "if" part, when $\phi(t) = \log(f_0(t))$ is log-concave, for any $\delta_0 < \delta_1$, and any $x < y$, we have

$$\log \frac{f_{\delta_1}(y)}{f_{\delta_0}(y)} - \log \frac{f_{\delta_1}(x)}{f_{\delta_0}(x)} = \phi(y - \delta_1) - \phi(y - \delta_0) - \phi(x - \delta_1) + \phi(x - \delta_0) \geq 0, \tag{5.8}$$

where the last inequality is a simple consequence of concavity (see Lemma 5.2 below). This proves the "if" part. □

Lemma 5.2 *Let ϕ be any concave function on $\mathbb{R}$. For any $x < y \in \mathbb{R}$, and $\delta > 0$, we have*

$$\phi(x) + \phi(y + \delta) \leq \phi(y) + \phi(x + \delta).$$

Proof Pick $\lambda = \delta/(y - x + \delta)$, by concavity of f, we have

$$\lambda\phi(x) + (1 - \lambda)\phi(y + \delta) \leq \phi(\lambda x + (1 - \lambda)(y + \delta)) = \phi(y), \tag{5.9}$$

and

$$(1 - \lambda)\phi(x) + \lambda\phi(y + \delta) \leq \phi((1 - \lambda)x + \lambda(y + \delta)) = \phi(x + \delta). \tag{5.10}$$

Summing up (5.9) and (5.10) and we arrive at the conclusion as desired. □

Theorem 5.1 and Lemma 5.1 yield immediately the following.

Corollary 5.1 *Consider the additive error model (1.1), where the $\epsilon(i)$'s are independent with common distribution F. Let the signal μ have s positive entries with magnitudes $0 < \delta_1 \leq \ldots \leq \delta_s$, located on $\{1, \ldots, p\}$ as prescribed in (5.2).*

If F has a positive, log-concave density f, then the support estimator

$$\widehat{S}^* := \{i \,:\, x(i) \geq x_{[s]}\}$$

is finite-sample optimal in terms of Bayes risk in the sense of (5.5).

Proof The independence and the fact that the observations have densities implies the absence of ties among the order statistics $\{x_{[i]}\}$, with probability one. Thus, the oracle thresholding procedure is a.s. unique and given by $\widehat{S}^* = \{i \,:\, x(i) \geq x_{[s]}\}$. The result then follows from Theorem 5.1 and Lemma 5.1.   □

Remark 5.1 Theorem 5.1 and Corollary 5.1 show that under MLR (or equivalently, log-concavity of the error densities in additive models), the oracle thresholding procedures are finite-sample optimal even in the case where the signals have different (positive) sizes. This fascinating property perhaps explains the success of the thresholding estimators.

The assumption of log-concavity of the densities is compatible with the AGG model when $\nu \geq 1$, as demonstrated in the next example.

Example 5.1 The generalized Gaussian density $f(x) \propto \exp\{-|x|^\nu/\nu\}$ is log-concave for all $\nu \geq 1$. Therefore in the additive error model (1.1), according to Corollary 5.1, the oracle thresholding procedure is Bayes optimal in the sense of (5.5).

5.3 Bayes Optimality of Likelihood Ratio Thresholding

When the MLR condition in Theorem 5.1 is violated, the oracle thresholding procedures can in fact be sub-optimal (see Example 5.2 and Sect. 5.4, below).

In this section, we demonstrate that thresholding the *likelihood ratio* rather than signal values yields the finite-sample Bayes optimal procedures. We consider a special but sufficiently general case of signal models with equal densities.

Namely, let the observations $x(i)$, $i = 1, \ldots, p$ have s signals as prescribed in (5.2) with *common* "signal" density f_a, and let the remaining $(p - s)$ locations have common "error" density f_0. Define the likelihood ratios

$$L(i) := f_a(x(i))/f_0(x(i)),$$

and let $L_{[1]} \geq L_{[2]} \geq \ldots \geq L_{[p]}$ be the order statistics of the $L(i)$'s.

Definition 5.3 (*Oracle likelihood ratio thresholding*) Recall that $\langle a_1, \cdots, a_s \rangle$ denotes the vector of a_i's arranged in a non-increasing order. Any estimator $\hat{S} = \{i_1, \cdots, i_s\}$ such that

$$\langle L(i_1), \cdots, L(i_s) \rangle = (L_{[1]}, \cdots, L_{[s]}),$$

will be referred to as an *oracle likelihood thresholding* estimator of the support S.

Theorem 5.2 *Any oracle likelihood ratio thresholding procedure* $\widehat{S}_{LRT}$ *is finite-sample optimal in terms of Bayes risk. That is,*

$$\widehat{S}_{LRT} \in \arg\min_{\widehat{S} \in \mathcal{S}} \mathbb{E}_\pi \mathbb{P}[\widehat{S} \neq S] \tag{5.11}$$

for all s and p, where the infimum on $\widehat{S}$ *is taken over all support estimators of size s.*

Proof The proof is analogous to that of Theorem 5.1. We need to show that $\widehat{S}_{LRT} \in$ arg max$_{S \in \mathcal{S}} f(x|S)\pi(S)$. Since the distribution π of the support S is uniform (recall (5.3)), it is equivalent to prove that

$$f(x|\widehat{S}_{LRT}) = \max_{S \subset \mathcal{S}} f(x|S),$$

where $f(x|S)$ is the conditional distribution of the data given the unordered support S,

$$f(x|S) = \sum_P f(x|P)\pi^{\text{ord}}(P|S) = \prod_{j \in S} f_a(x(j)) \prod_{j \notin S} f_0(x(j)). \tag{5.12}$$

Suppose $\widehat{S} \in \mathcal{S}$ is *not* an oracle likelihood thresholding estimator. Then from the definition of the likelihood ratio thresholding procedure, there must be indices $j \in \widehat{S}$ and $j' \notin \widehat{S}$ such that $L(j) < L(j')$. If we exchange the labels of $L(j)$ and $L(j')$, that is, we form a new estimate $\widehat{S}' = (\widehat{S} \setminus \{j\}) \cup \{j'\}$, comparing the log-likelihoods under $\widehat{S}$ and $\widehat{S}'$, we have

$$\log f(x|\widehat{S}) - \log f(x|\widehat{S}') = \log f_a(x(j)) + \log f_0(x(j')) - \log f_a(x(j')) - \log f_0(x(j)).$$

By the definition of $L(j)$'s, and the order relations, we obtain

$$\log f(x|\widehat{S}) - \log f(x|\widehat{S}') = \log L(j) - \log L(j') > 0$$

This shows that $\widehat{S}$ cannot be Bayes optimal unless it is a likelihood thresholding estimator. Note that with the same argument for every two likelihood thresholding estimators $\widehat{S}'$ and $\widehat{S}''$, we have $f(x|\widehat{S}') = f(x|\widehat{S}'')$, proving the desired optimality. $\square$

The characterization of optimal likelihood ratio thresholding procedures in Theorem 5.2 may not always yield practical estimators, as the density of the alternatives, and the number of signals s are typically unknown. Still, some insights can be gained by virtue of Theorem 5.2. In particular, when MLR fails (for example, when the errors in model (1.1) do not have log-concave densities), data thresholding is sub-optimal.

Example 5.2 (*Sub-optimality of data thresholding*) Let the errors have iid generalized Gaussian density with $\nu = 1/2$, i.e., $\log f_0(x) \propto -x^{1/2}$. Let dimension $p = 2$, sparsity $s = 1$ with uniform prior, and signal size $\delta = 1$. That is, $\mathbb{P}[\mu = (0, 1)^T] = \mathbb{P}[\mu = (1, 0)^T] = 1/2$. If the observations take on values $x = (x_1, x_2)^T = (1, 2)^T$, we see from a comparison of the likelihoods (and hence, the posteriors),

$$\log \frac{f(x|\{1\})}{f(x|\{2\})} = 2x_1^{1/2} + 2(x_2 - 1)^{1/2} - 2x_2^{1/2} - 2(x_1 - 1)^{1/2} = 4 - 2\sqrt{2} > 0,$$

that even though $x_1 < x_2$, the set $\{1\}$ is a better estimate of support than $\{2\}$, i.e., $\mathbb{P}[S = \{1\} \mid x] > \mathbb{P}[S = \{2\} \mid x]$.

This simple example shows that, in the case when the errors have super-exponential tails, the optimal procedures are in general *not* data thresholding. A slightly more general conclusion can be found in Corollary 5.2.

5.4 Sub-optimality of Data Thresholding Procedures

We provide a slightly more general result on the sub-optimality of data thresholding procedures.

Corollary 5.2 *Consider the additive error model (1.1). Let the errors ϵ be independent with common distribution F. Let each of the s signals be located on $\{1, \ldots, p\}$ uniformly at random with equal magnitude $0 < \delta < \infty$. Assume the errors $\epsilon(i)$'s are iid with density f that is log-convex on $[K, +\infty)$, for some $K > 0$.*

If $\widehat{S}_{opt}$ is the Bayes optimal (i.e., the oracle likelihood thresholding estimator), then, whenever $j \in \widehat{S}_{opt}$ for some $x(j) > K + \delta$, we must necessarily have $j' \in \widehat{S}_{opt}$ for all j' such that $K + \delta \leq x(j') < x(j)$.

Specifically, if there are m observations exceeding $K + \delta$, with $m > s$, then the top $m - s$ observations will *not* be included in the optimal estimator $\widehat{S}_{opt}$. This shows that, in the case when the errors have super-exponential tails, the optimal procedures are in general *not* data thresholding.

Proof (*Corollary 5.2*) Since the density of the alternatives $f_a(t) = f(t - \delta)$ is log-convex on $[K + \delta, \infty)$, by Relation (5.8) in the proof of Lemma 5.1 and appealing to log-convexity (rather than log-concavity), the likelihood ratio

$$L(j) := \frac{f_a(x(j))}{f_0(x(j))}$$

is decreasing in $x(j)$ on $[K + \delta, \infty)$. The claim follows from Theorem 5.2. $\qquad\square$

Remark 5.2 As we have seen, the thresholding estimators are no longer optimal in the additive model with error densities heavier than exponential. Thanks to Theorem 5.2, the oracle likelihood thresholding procedures are promising alternatives that can lead us to practical support estimators.

In the case where the signals have different sizes, however, the argument in the proof of Theorem 5.2 breaks downs since the signal densities need to be identical for Relation (5.12) to hold. In such cases, the characterization of the optimal procedure is an open problem.

5.5 Minimax Optimality in Exact Support Recovery

We establish in this section minimax versions of our results from Chap. 4. Specifically, if we restrict ourselves to *the class of thresholding procedures* $\mathcal{T}$ (defined in (2.20)), then Bonferroni's procedure is minimax optimal, for *any* fixed dependence structure in the URS class. This is formalized in Corollary 5.3 below. We refer to this result as *point-wise* minimax, to emphasize the fact that this optimality holds for every *fixed* URS array.

Meanwhile, if we search over *all procedures*, but expand the model space to include *all* dependence structures, then a different minimax optimality statement holds for Bonferroni's procedure. This result, formally stated in Sect. 5.5.2, is a consequence of our characterization of the finite-sample Bayes optimality of thresholding procedures in Sect. 5.2.

5.5.1 Point-Wise Minimax Optimality for Thresholding Procedures

Theorems 4.1 and 4.2 can be cast in the form of an asymptotic minimax statement.

Corollary 5.3 (Point-wise minimax) *Let $\widehat{S}^{Bonf}$ be the sequence of Bonferroni's procedure described in Theorem 4.1. Let also the errors have common AGG(v) distribution F with parameter $v > 0$, and $\Theta_p^+(\beta, \underline{r})$ be as defined in (4.7). If $\underline{r} > f_E(\beta)$, then we have*

$$\limsup_{p \to \infty} \ \sup_{\mu \in \Theta_p^+(\beta, \underline{r})} \mathbb{P}(\widehat{S}_p^{Bonf} \neq S_p) = 0, \tag{5.13}$$

for arbitrary dependence structure of the error array $\mathcal{E} = \{\epsilon_p(i)\}_p$. Let $\mathcal{T}$ be the class of thresholding procedures (2.20). If $\underline{r} < f_E(\beta)$, then we have

$$\liminf_{p \to \infty} \ \inf_{\widehat{S}_p \in \mathcal{T}} \ \sup_{\mu \in \Theta_p^+(\beta, \underline{r})} \mathbb{P}(\widehat{S}_p \neq S_p) = 1, \tag{5.14}$$

for any error-dependence structure such that $\mathcal{E} \in U(F)$ and $(-\mathcal{E}) \in U(F)$.

Proof The first conclusion (5.13) is a restatement of Theorem 4.1.

For the second statement (5.14), since $\underline{r} < f_E(\beta)$, we can pick a sequence $\mu^* \in \Theta_p^+(\beta, \underline{r})$ such that $|S_p| = \lfloor \ell(p) p^{1-\beta} \rfloor$, with signals having the same signal size $\mu(i) = (2r \log p)^{1/\nu}$ for all $i \in S_p$, where $\underline{r} < r < f_E(\beta)$. For this particular choice of μ^*, we have $\mu^* \in \Theta_p^-(\beta, \bar{r})$ (recall (4.22)), where $r < \bar{r} < f_E(\beta)$, and according to Theorem 4.2, we obtain $\lim_{p \to \infty} \inf_{\widehat{S}_p \in \mathcal{T}} \mathbb{P}[\widehat{S}_p \neq S_p] = 1$, for all dependence structures in the URS class. $\square$

Remark 5.3 Theorem 4.2 is a stronger result than the traditional minimax claim in Relation (5.14). Indeed, (4.24) involves an infimum (over the class Θ_p^-), while (5.14) has a supremum (over the class Θ_p^+).

On the other hand, Corollary 5.3 is more informative than many minimax-type statements, since it applies "point-wise" to any fixed error-dependence structure in the URS class.

Remark 5.4 Corollary 5.3 echoes Remark 4.5: for a very large class of dependence structures, we cannot improve upon Bonferroni's procedure in exact support recovery problems (asymptotically), unless we look beyond thresholding procedures.

5.5.2 Minimax Optimality over All Procedures

Consider the asymptotic Bayes risk as defined in (5.4). The statement for the necessary condition of support recovery in Theorem 4.2, with the help of Corollary 5.1, can be strengthened to include all procedures (in the Bayesian context), regardless of whether they are thresholding.

Theorem 5.3 *Consider the additive model (1.1) where the $\epsilon_p(i)$'s are independent and identically distributed with log-concave densities in the AGG class. Let the signals be as prescribed in Corollary 5.1. If the signal sizes fall below the strong classification boundary (4.5), i.e. $\bar{r} < f_E(\beta)$, then we have*

$$\liminf_{p \to \infty} \inf_{\widehat{S}_p} \mathbb{E}_\pi \mathbb{P}[\widehat{S}_p \neq S_p] = 1, \tag{5.15}$$

where the infimum on $\widehat{S}_p$ is taken over all procedures.

Proof When the errors are independent with log-concave density, the oracle thresholding procedure $\widehat{S}_p^*$, by Corollary 5.1, minimizes the Bayes risk (5.4) among *all* procedures. That is,

$$\liminf_{p \to \infty} \inf_{\widehat{S}_p} \mathbb{E}_\pi \mathbb{P}[\widehat{S}_p \neq S_p] \geq \liminf_{p \to \infty} \mathbb{E}_\pi \mathbb{P}[\widehat{S}_p^* \neq S_p].$$

Since $\widehat{S}_p^*$ belongs to the class of all thresholding procedures, we have

$$\liminf_{p \to \infty} \mathbb{E}_{\pi} \mathbb{P}[\widehat{S}_p^* \neq S_p] \geq \liminf_{p \to \infty} \inf_{\widehat{S}_p \in \mathcal{T}} \mathbb{E}_{\pi} \mathbb{P}[\widehat{S}_p \neq S_p]$$

$$\geq \liminf_{p \to \infty} \inf_{\widehat{S}_p \in \mathcal{T}} \inf_{S_p} \mathbb{P}[\widehat{S}_p \neq S_p] = 1,$$

when $\overline{r} < f_E(\beta)$, where the last line follows from Theorem 4.2. $\qquad\square$

Theorem 5.3 allows us to state another minimax conclusion—one in which we search over *all procedures*, by allowing the supremum in the minimax statement to be taken over the dependence structures.

Corollary 5.4 *Let $D(F)$ be the collection of error arrays with common marginal F as defined in (4.20) where F is an $AGG(v)$ distribution. Let also $\widehat{S}_p^{Bonf}$ be Bonferroni's procedure as described in Theorem 4.1. If $\underline{r} > f_E(\beta)$, then we have*

$$\limsup_{\substack{p \to \infty \\ \mu \in \Theta_p^+(\beta, \underline{r}) \\ \mathcal{E} \in D(F)}} \sup \; \mathbb{P}(\widehat{S}_p^{Bonf} \neq S_p) = 0. \tag{5.16}$$

Further, when $\underline{r} < f_E(\beta)$, and F has a positive log-concave density f, we have

$$\liminf_{p \to \infty} \inf_{\widehat{S}_p} \sup_{\substack{\mu \in \Theta_p^+(\beta, \underline{r}) \\ \mathcal{E} \in D(F)}} \mathbb{P}(\widehat{S}_p \neq S_p) = 1, \tag{5.17}$$

where the infimum on $\widehat{S}_p$ is taken over all procedures.

Proof Relation (5.16) is a re-statement of Remark 4.1.

For any distribution π (with a slight abuse of notation) over the parameter space $\Theta_p^+ \times D(F)$, we have

$$\liminf_{p \to \infty} \inf_{\widehat{S}_p} \sup_{\substack{\mu \in \Theta_p^+(\beta, \underline{r}) \\ \mathcal{E} \in D(F)}} \mathbb{P}(\widehat{S}_p \neq S_p) \geq \liminf_{p \to \infty} \inf_{\widehat{S}_p} \mathbb{E}_{\pi} \mathbb{P}(\widehat{S}_p \neq S_p), \tag{5.18}$$

since the supremum is bounded from below by expectations. In particular, define π to be the uniform distribution over the configurations $\Theta_p^* \times I(f)$, where

$$\Theta_p^* = \{\mu \in \mathbb{R}^d : |S_p| = \lfloor \ell(p) p^{1-\beta} \rfloor, \; \mu(i) = 0 \text{ for all } i \notin S, \text{ and}$$

$$\mu(i) = (vr \log p)^{1/v} \text{ for all } i \in S, \text{ where } \underline{r} < r < f_E(\beta)\},$$

and

$$I(f) = \{\mathcal{E} = (\epsilon_p(i))_p : \epsilon_p(i) \text{ iid with density } f(x) \propto \exp\{-|x|^v / v\}\}.$$

Since the density f of F is log-concave, the distribution of the signal configurations satisfies the conditions in Theorem 5.3. Thus, the desired conclusion (5.17) follows from Theorem 5.3 and (5.18). $\qquad\square$

Remark 5.5 Since the class AGG(v), $v \geq 1$ contains distributions with log-concave densities (Example 5.1), the minimax statement (5.17) continues to hold if the supremum is taken over the entire class $F \in \text{AGG}(v)$, $v \geq 1$. We opted for a more informative formulation which emphasizes the log-concavity condition on the density of F.

Remark 5.6 Corollary 5.4 is no stronger than Corollary 5.3. In Corollary 5.3 we search over only the class of thresholding procedures, but offer a tight, point-wise lower bound on the asymptotic risk over the class of URS-dependence structures. On the other hand, Corollary 5.4 provides a uniform lower bound for the asymptotic risk over all dependence structures, which may not be tight except in the case of independent errors.

5.6 Optimality and Sub-optimality: A Discussion

We conclude with a brief summary on the optimality and sub-optimality of the thresholding procedures in the problem of exact support estimation. For clarity, we focus on the model (1.1) with *independent* errors.

Theorem 5.3 and Corollary 5.4 provide a nearly complete picture of the difficulty in the exact support recovery problem, in the regime when the thresholding estimators are optimal. Specifically, in such cases the signal classification boundary is universal. On the other hand, Theorem 5.2, and indeed, Example 5.2 demonstrate that thresholding procedures are *sub-optimal* for AGG(v) models with $v < 1$. Therefore, the optimality of thresholding procedures (specifically, Bonferroni's procedure) only applies to AGG(v) models with $v \geq 1$.

If we restrict the space of methods to only thresholding procedures, then the results in Sect. 5.5.1 state that the phase transition phenomenon—the 0–1 law in the sense of Corollary 5.3—is universal in all error models with rapidly varying tails. This includes AGG(v) models *for all* $v > 0$. In contrast, models with heavy (regularly varying) tailed errors do not exhibit this phenomenon (form more details, see Theorem B.3). We summarize the properties of thresholding procedures in Table 5.1.

Table 5.1 Properties of thresholding procedures under different error distributions when the errors are independent. Properties of the error distributions are listed in brackets

Thresholding procedure (Error distributions)	Bayes optimality (Log-concave density)	Phase transition (Rapidly-varying tails)
AGG(v), $v \geq 1$	Yes (Yes)	Yes (Yes)
AGG(v), $0 < v < 1$	No (No)	Yes (Yes)
Power laws	No (No)	No (No)

Chapter 6
Uniform Relative Stability for Gaussian Arrays

The notion of uniform relative stability (URS) in Definition 4.1 is the key to the necessary conditions for exact support recovery established in Theorem 4.2. In this chapter, we provide a complete characterization of the class of URS Gaussian arrays in terms of a simple condition on their covariance structure. The condition is as follows.

Definition 6.1 (*Uniformly decreasing dependence (UDD)*) Consider a triangular array of jointly Gaussian distributed errors $\mathcal{E} = \left\{ \left(\epsilon_p(i) \right)_{i=1}^{p}, p = 1, 2, \dots \right\}$ with unit variances,

$$\epsilon_p \sim N(0, \Sigma_p), \quad p = 1, 2, \dots.$$

The array $\mathcal{E}$ is said to be uniform decreasingly dependent (UDD) if for every $\delta > 0$ there exists a finite $N(\delta) < \infty$, such that for every $i \in \{1, \dots, p\}$, and $p \in \mathbb{N}$, we have

$$\left| \left\{ k \in \{1, \dots, p\} : \Sigma_p(i, k) > \delta \right\} \right| \le N(\delta) \quad \text{for all } \delta > 0. \tag{6.1}$$

That is, for every coordinate i, the number of elements which are more than δ-correlated with $\epsilon_p(i)$ does not exceed $N(\delta)$.

Note that the bound in (6.1) holds uniformly in i and p, and only depends on δ. Also observe that on the left-hand side of (6.1), we merely count in each row of Σ_p the number of exceedances of covariances (not their absolute values!) over level δ.

Remark 6.1 Without loss of generality, we may require that $N(\delta)$ be a monotone non-increasing function of δ, for we can take

$$N(\delta) = \sup_{p,i} \left| \{k : \Sigma_p(i, k) > \delta\} \right|,$$

which is non-increasing in δ. Definition 6.1 therefore states that the array is UDD when $N(\delta) < \infty$ for all $\delta > 0$.

© The Author(s), under exclusive license to Springer Nature Switzerland AG 2021   75
Z. Gao and S. Stoev, *Concentration of Maxima and Fundamental
Limits in High-Dimensional Testing and Inference*,
SpringerBriefs in Probability and Mathematical Statistics,
https://doi.org/10.1007/978-3-030-80964-5_6

Observe that the UDD condition does not depend on the order of the coordinates in the error vector $\epsilon_p = (\epsilon_p(i))_{i=1}^p$. Often times, however, the errors are thought of as coming from a stochastic process indexed by time or space. To illustrate the generality of the UDD condition, we formulate next a simple sufficient condition (UDD') that is easier to check in a time-series context.

Definition 6.2 *(UDD')* For $\epsilon_p \sim N(0, \Sigma_p)$ with unit variances, an array $\mathcal{E} = (\epsilon_p(i))_{i=1}^p$ is said to satisfy the UDD' condition if there exist:

(i) permutations l_p of $\{1, \ldots, p\}$, for all $p \in \mathbb{N}$, and
(ii) a non-negative sequence $(r_n)_{n=1}^\infty$ converging to zero $r_n \to 0$, as $n \to \infty$,

such that

$$\sup_{p \in \mathbb{N}} |\Sigma_p (i', j')| \le r_{|i-j|}. \tag{6.2}$$

where $i' = l_p(i)$, $j' = l_p(j)$, for all $i, j \in \{1, \ldots, p\}$.

Remark 6.2 Without loss of generality, we may also require that r_n be non-increasing in n, for we can replace r_n with the non-increasing sequence $r_n' = \sup_{m \ge n} r_m$.

Proposition 6.1 *UDD' implies UDD.*

Proof Since $r_n \to 0$, for any $\delta > 0$, there exists an integer $M = M(\delta) < \infty$ such that $r_n \le \delta$, for all $n \ge M$. Thus, by (6.2), for every fixed $j' \in \{1, \ldots, p\}$, we can have $|\mathrm{Cov}(\epsilon_p(k'), \epsilon_p(j'))| > \delta$, only if k' belongs to the set:

$$\{k' \in \{1, \ldots, p\} : j - M \le k := l_p^{-1}(k') \le j + M\},$$

where $j := l_p^{-1}(j')$. That is, there are at most $2M + 1 < \infty$ indices $k' \in \{1, \ldots, p\}$, whose covariances with $\epsilon(j')$ may exceed δ. Since this holds uniformly in $j' \in \{1, \ldots, p\}$, Condition UDD follows with $N(\delta) = 2M + 1$. $\qquad\square$

We now state the main result of this chapter. It states that a Gaussian array is URS if and only if it is UDD. The URS condition essentially requires that the dependencies decay in a uniform fashion, the rate at which dependence decay does *not* matter.

Theorem 6.1 *Let $\mathcal{E}$ be a Gaussian triangular array with standard normal marginals. The array $\mathcal{E}$ has uniformly relatively stable (URS) maxima if and only if it is uniformly decreasing dependent (UDD).*

Specifically, for stationary Gaussian arrays, we have the following corollary.

Corollary 6.1 *Let $\mathcal{E} = \{\epsilon_p(i) = Z(i)\}$ for a stationary Gaussian time series $\mathcal{Z} = \{Z(i)\}$. Then $\mathcal{E}$ is URS if and only if the autocovariance function $\mathrm{Cov}(Z(k), Z(0)) \to 0$, as $k \to \infty$.*

Corollary 6.1 follows by Theorem 6.1 and the observation that UDD is equivalent to vanishing autocovariance of $\mathcal{Z}$. A slightly weaker form of the "if" part was established in Theorem 3 of Berman (1964).

Returning again to the study of support recovery problems, Theorem 6.1 and the necessary condition for exact support recovery in Theorem 4.2 yield the following result.

Corollary 6.2 *For UDD Gaussian errors, the result in Theorem 4.2 holds.*

One may ask, whether the UDD (equivalently, URS) condition can be relaxed further for the phase-transition result in Theorem 4.2 to hold. As a counterpart to Remark 4.5, we demonstrate next that the dependence conditions in Theorem 4.2 are nearly optimal. Specifically, we show that if the URS-dependence condition is violated, then it may be possible to recover the support of weaker signals, falling below the boundary. The main idea is to use the equivalence of URS and UDD to construct a Gaussian error array, whose correlations do not decay in a uniform fashion (UDD fails). As we will see, in such a case one can do substantially better in terms of support recovery. This shows that the URS condition is nearly optimal in the Gaussian setting. Numerical simulations illustrating this example can be found in Sect. 4.6, below.

Example 6.1 (*On the tightness of the URS condition for exact support recovery*)
Suppose $\mathcal{E} = (\epsilon_p(i))_{i=1}^p$ is Gaussian, and is comprised of $\lfloor p^{1-\beta} \rfloor$ blocks, each of size at least $\lfloor p^\beta \rfloor$. Let the elements within each block have correlation 1, and let the elements from different blocks be independent. If $\underline{r \geq 4(1 - \beta)}$, then the procedure

$$\widehat{S} = \left\{ i : x(i) > \sqrt{2(1 - \beta) \log p} \right\}$$

yields exact support recovery, i.e., $\mathbb{P}[\widehat{S} = S] \to 1$, as $p \to \infty$. This requirement on the signal size is strictly weaker than that of the strong classification boundary, since $4(1 - \beta) < (1 + \sqrt{1 - \beta})^2$ on $\beta \in (0, 1)$.

Proof (*Example 6.1*) Let $t_p^* = \sqrt{2(1 - \beta) \log p}$ and observe that $\widehat{S} = \{ j : x(j) > t_p^* \}$. Analogous to (4.11) in the proof of Theorem 4.1, we have

$$\mathbb{P}\left[\widehat{S} \subseteq S \right] = 1 - \mathbb{P}\left[\max_{j \in S^c} x(j) > t_p^* \right] = 1 - \mathbb{P}\left[\max_{j \in S^c} \epsilon(j) > t_p^* \right]$$

$$\geq 1 - \mathbb{P}\left[\max_{j \in \{1, \dots, p\}} \epsilon(j) > t_p^* \right] \geq 1 - \mathbb{P}\left[\max_{j \in \{1, \dots, \lfloor p^{1-\beta} \rfloor\}} \widetilde{\epsilon}(j) > t_p^* \right]$$

where $(\widetilde{\epsilon})_{j=1}^{\lfloor p^{1-\beta} \rfloor}$'s are independent Gaussian errors; in the last inequality we used the assumption that there are at most $\lfloor p^{1-\beta} \rfloor$ independently distributed Gaussian errors in $(\epsilon_p(j))_{j=1}^p$. By Example 4.1 (with $\lfloor p^{1-\beta} \rfloor$ taking the role of p), we know that the FWER goes to 0 at a rate of $(2 \log \lfloor p^{1-\beta} \rfloor)^{-1/2}$. Therefore, the probability of no false inclusion converges to 1.

On the other hand, since the signal sizes are no smaller than $(\nu \underline{r} \log p)^{1/\nu} = \sqrt{2\underline{r} \log p}$ (for $\nu = 2$), similar to (4.13), we obtain

$$
\begin{aligned}
\mathbb{P}\big[\widehat{S} \supseteq S\big] &\geq \mathbb{P}\left[\min_{j \in S} \epsilon(j) > \sqrt{2(1-\beta)\log p} - \sqrt{2\underline{r}\log p}\right] \\
&= \mathbb{P}\left[\max_{j \in S}(-\epsilon(j)) < \sqrt{2\log p}\left(\sqrt{\underline{r}} - \sqrt{1-\beta}\right)\right] \\
&= \mathbb{P}\left[\frac{\max_{j \in S}(-\epsilon(j))}{u_{|S|}} < \frac{\sqrt{\underline{r}} - \sqrt{1-\beta}}{\sqrt{1-\beta}}(1 + o(1))\right], \quad (6.3)
\end{aligned}
$$

where in the last line we used the quantiles (2.33). Since the minimum signal size is bounded below by $\underline{r} > 4(1 - \beta)$, the right-hand side of the inequality in (6.3) converges to a constant strictly larger than 1. While the left-hand side, by Slepian's lemma (recall Theorem 2.1 and Relation 2.47), is stochastically smaller than a r.v. going to 1. Namely, we have

$$
\frac{1}{u_{|S|}}\max_{j \in S}(-\epsilon(j)) \overset{d}{\leq} \frac{1}{u_{|S|}}\max_{j \in S}\epsilon^*(j) \overset{\mathbb{P}}{\longrightarrow} 1, \quad (6.4)
$$

where $(\epsilon^*)_{j=1}^{\lfloor p^{1-\beta}\rfloor}$'s are independent Gaussian errors. Therefore the probability in (6.3) must also converge to 1. $\qquad\square$

Before proceeding to the proof of Theorem 6.1, we will briefly discuss the relationships between UDD and other dependence conditions in the context of extreme value theory. The main idea we would like to convey is that UDD (and equivalently URS) is an exceptionally mild condition on the dependence of the array.

The Berman and UDD conditions. Suppose that the array of errors $\mathcal{E}$ comes from a stationary Gaussian time series $\epsilon(i)$, $i \in \mathbb{N}$, with autocovariance $r_p = \mathrm{Cov}(\epsilon(i + p), \epsilon(i))$. One is interested in the asymptotic behavior of the maxima $M_p := \max_{i=1,\dots,p} \epsilon(i)$.

In this setting, Berman's condition, introduced in Berman (1964), requires that

$$
r_p \log p \to 0, \quad \text{as } p \to \infty. \quad (6.5)
$$

This condition entails that

$$
a_p(M_p - b_p) \overset{d}{\longrightarrow} Z, \quad \text{as } p \to \infty, \quad (6.6)
$$

with the Gumbel limit distribution $\mathbb{P}[Z \leq x] = \exp\{-e^{-x}\}$, $x \in \mathbb{R}$, where

$$
a_p = \sqrt{2\log p}, \quad b_p = \sqrt{2\log p} - \frac{1}{2}\left(\sqrt{2\log p}\right)^{-1}(\log\log(p) + \log(4\pi)),
$$

are *the same* centering and normalization sequences as in the case of iid $\epsilon(i)$'s. Berman's condition is one of the weakest dependence conditions in the literature for which the convergence in (6.6) holds. See, e.g., Theorem 4.4.8 in Embrechts et al. (2013), where (6.5) is described as "very weak".

Instances where the dependence in the time series is so strong that Berman's condition (6.5) fails have also been studied. In such cases, one may continue to have (6.6) but typically the sequences of normalizing and centering constants will be *different* from the iid case, and the corresponding limit is usually no longer Gumbel; see, for example, Theorems 6.5.1 and 6.6.4 in Leadbetter et al. (1983) and McCormick and Mittal (1976).

In our high-dimensional support estimation context, the notion of relative stability is sufficient and more natural than the finer notions of distributional convergence. If one is merely interested in the asymptotic relative stability of the Gaussian maxima, then Berman's condition can be relaxed significantly (see also, Theorem 4.1 of Berman 1964). Observe that by Proposition 6.1, the Berman condition (6.5) implies UDD and hence relative stability (Theorem 6.1), i.e.,

$$\frac{1}{b_p} M_p \xrightarrow{\mathbb{P}} 1, \quad \text{as} \quad p \to \infty. \tag{6.7}$$

This *concentration of maxima* property can be readily deduced from (6.6), since $a_p b_p \sim 2 \log(p) \to \infty$ as $p \to \infty$. Theorem 6.1 shows that (6.7) holds if the much weaker uniform dependence condition UDD holds. Note that our condition is coordinate free—neither monotonicity of the sequence r_p nor stationarity of the underlying array is required. This makes it substantially broader than the time series setting in the seminal work Berman (1964).

The rest of this chapter is devoted to the proof of the main result, i.e., Theorem 6.1. We first introduce a key lemma regarding the structure of an *arbitrary* correlation matrix of high-dimensional random variables. The proof uses a surprising, yet elegant application of Ramsey's Theorem from the study of combinatorics. The "only if" part of Theorem 6.1 follows from this lemma, in Sect. 6.2.

The proof of the "if" part is detailed in Sect. 6.3. The arguments there have been recently extended to establish bounds on the rate of concentration of maxima in Kartsioukas et al. (2019); see also, Tanguy (2015b) and the related notion of super-concentration of Chatterjee (2014).

6.1 Ramsey's Theory and the Structure of Correlation Matrices

Given any integer $k \geq 1$, there is always an integer $R(k, k)$ called the *Ramsey number*:

$$k \leq R(k, k) \leq \binom{2k - 2}{k - 1} \tag{6.8}$$

such that the following property holds: every undirected graph with at least $R(k, k)$ vertices will contain *either* a clique of size k, or an *independent set* of k nodes. Recall that a clique is a complete sub-graph where all pairs of nodes are connected, and an independent set is a set of nodes where no two nodes are connected.

This result is a consequence of the celebrated work of Ramsey (2009), which gave birth to Ramsey Theory (see e.g., Conlon et al. 2015). The Ramsey Theorem and the upper bound (6.8) (established first in Erdös and Szekeres 1935) are at the heart of the proof of the following result. A recent improvement on the upper bound is given by Sah (2020).

Proposition 6.2 *Fix $\gamma \in (0, 1)$ and let $P = (\rho(i, j))_{n \times n}$ be an arbitrary correlation matrix. If*

$$k := \lfloor \log_2(n)/2 \rfloor \geq \lceil 1/\gamma \rceil + 1, \tag{6.9}$$

then there is a set of k indices $K = \{l_1, \ldots, l_k\} \subseteq \{1, \ldots, n\}$ such that

$$\rho(i, j) \geq -\gamma, \text{ for all } i, j \in K. \tag{6.10}$$

Proof By using (6.8) and a refinement of Stirling's formula, we will show at the end of the proof that for $k \leq \log_2(n)/2$, we have

$$R(k, k) \leq n, \tag{6.11}$$

where $R(k, k)$ is the Ramsey number.

Now, construct a graph with vertices $\{1, \ldots, n\}$ such that there is an edge between nodes i and j if and only if $\rho(i, j) \geq -\gamma$. In view of (6.11) and Ramsey's theorem (see e.g., Theorem 1 in Fox (2009) or Conlon et al. (2015) for a recent survey on Ramsey theory), there is a subset of k nodes $K = \{l_1, \ldots, l_k\}$, which is either a *complete graph* or an *independent set*. Recall that in a complete graph, every two nodes are connected with an edge; while in independent sets, no two nodes are connected.

If K is a complete graph, then by our construction of the graph, Relation (6.10) holds.

Now, suppose that K is a set of independent nodes. This means, again by the construction of our graph, that

$$\rho(i, j) < -\gamma, \quad \text{for all } i \neq j \in K.$$

Let Z_i, $i \in K$ be zero-mean random variables such that $\rho(i, j) = \mathbb{E}[Z_i Z_j]$. Observe that

$$\text{Var}\left(\sum_{i \in K} Z_i\right) = \sum_{i \in K} \text{Var}(Z_i) + \sum_{\substack{i \neq j \\ i, j \in K}} \text{Cov}(Z_i, Z_j) < k - k(k - 1)\gamma, \tag{6.12}$$

since $\text{Var}(Z_i) = 1$ and $\rho(i, j) < -\gamma$ for $i \neq j$. By our assumption, $k \geq (\lceil 1/\gamma \rceil + 1)$, or equivalently, $(k - 1) \geq 1/\gamma$, the variance in (6.12) is negative. This is a contradiction showing that there are no independent sets K with cardinality k.

To complete the proof, it remains to show that Relation (6.11) holds. In view of the upper bound on the Ramsey numbers (6.8), it is enough to show that $k \leq \log_2(\sqrt{n})$ implies

$$\binom{2k - 2}{k - 1} \leq n.$$

This follows from a refinement of the Stirling formula, due to Robbins (1955):

$$\sqrt{2\pi} m^{m+1/2} e^{-m} e^{\frac{1}{(12m+1)}} \leq m! \leq \sqrt{2\pi} m^{m+1/2} e^{-m} e^{\frac{1}{12m}}.$$

Indeed, letting $\widetilde{k} := k - 1$, and applying the above upper and lower bounds to the terms $(2\widetilde{k})!$ and $\widetilde{k}!$, respectively, we obtain

$$\binom{2k - 2}{k - 1} \equiv \frac{(2\widetilde{k})!}{(\widetilde{k}!)^2} \leq \frac{2^{2\widetilde{k}}}{\sqrt{\pi \widetilde{k}}} \exp\left\{\frac{1}{24\widetilde{k}} - \frac{2}{12\widetilde{k} + 1}\right\} < 2^{2k}$$

where the last two inequalities follow by simply dropping positive factors less than 1. Since $2k \leq \log_2(n)$, the above bound implies Relation (6.11) and the proof is complete. $\qquad\square$

Using Proposition 6.2, we establish the key lemma used in the proof of Theorem 6.1.

Lemma 6.1 *Let $c \in (0, 1)$, and $P = (\rho(i, j))_{(n+1) \times (n+1)}$ be a correlation matrix such that*

$$\rho(1, j) > c \quad \text{for all } j = 1, \ldots, n + 1. \tag{6.13}$$

If $n \geq 2^{2\lceil 2/c^2 \rceil + 4}$, then there is a set of indices $K = \{l_1, \ldots, l_k\} \subseteq \{2, \ldots, n + 1\}$ of cardinality $k = |K| = \lfloor \log_2 \sqrt{n} \rfloor$, such that

$$\rho(i, j) > \frac{c^2}{2} \quad \text{for all } i, j \in K. \tag{6.14}$$

That is, all entries of the $k \times k$ sub-correlation matrix $P_K := (\rho(i, j))_{i,j \in K}$ are larger than $c^2/2$.

Proof *(Lemma 6.1)* Let $Z_1, \ldots, Z_{n+1}$ be random variables with covariance matrix P. Denote $\rho_j = \rho(1, j)$ and define

$$R_j = \begin{cases} \frac{1}{\sqrt{1-\rho_j^2}} (Z_j - \rho_j Z_1), & \text{if } \rho_j < 1, \\ R^* & \text{if } \rho_j = 1, \end{cases} \tag{6.15}$$

where R^* is an arbitrary zero-mean, unit-variance random variable. It is easy to see that $\mathrm{Var}(R_j) = 1$, and

$$\mathrm{Cov}\left(Z_i, Z_j\right) = \mathrm{Cov}\left(\rho_i Z_1 + \sqrt{1 - \rho_i^2} R_i, \ \rho_j Z_1 + \sqrt{1 - \rho_j^2} R_j\right)$$

$$= \rho_i \rho_j + \sqrt{1 - \rho_i^2}\sqrt{1 - \rho_j^2}\, \mathrm{Cov}\left(R_i, R_j\right)$$

$$> c^2 + \min\left\{\mathrm{Cov}\left(R_i, R_j\right), 0\right\}.$$

Therefore, Relation (6.14) would hold if we can find a set of indices $K = \{l_1, \ldots, l_k\}$ such that $\mathrm{Cov}\left(R_i, R_j\right) \geq -c^2/2$ for all $i, j \in K$, where $k = |K| = \lfloor \log_2 \sqrt{n} \rfloor$. This, however, follows from Proposition 6.2 applied to $(R_j)_{j=2}^{n+1}$ with $\gamma = c^2/2$, provided that

$$k = \lfloor \log_2 \sqrt{n} \rfloor \geq \lceil 2/c^2 \rceil + 1.$$

The last inequality indeed follows form the assumption that $n \geq 2^{2\lceil 2/c^2 \rceil + 4}$. $\qquad\square$

6.2   URS Implies UDD (Proof of the "Only If" Part of Theorem 6.1)

In view of Remark 6.1, UDD is equivalent to the requirement that $N(\delta) := 1 + \sup_p N_p(\delta) < \infty$ for all $\delta \in (0, 1)$, where

$$N_p(\delta) := \max_{j \in \{1, \ldots, p\}} \left| \{i : i \neq j, \ \Sigma_p(j, i) > \delta\} \right|. \tag{6.16}$$

Therefore, if $\mathcal{E}$ is not UDD, then there must exist a constant $c \in (0, 1)$ for which $N(c)$ is infinite, i.e., there is a subsequence $\widetilde{p} \to \infty$ such that $N_{\widetilde{p}}(c) \to \infty$. Without loss of generality, we may assume that $\widetilde{p} = p$.

Let $j_p(c)$ be the maximizers of (6.16), and let

$$S_p(c) := \{i \in \{1, \ldots, p\} : \Sigma_p(j_p(c), i) > c\}. \tag{6.17}$$

Observe that $|S_p(c)| = N_p(c) + 1 \to \infty$, as $p \to \infty$ (note that $j_p(c) \in S_p(c)$).

Applying Lemma 6.1 to the set of random variables indexed by $S_p(c)$, we conclude, for $N_p(c) \geq 2^{2\lceil 2/c^2 \rceil + 4}$, there must be a further subset

$$K_p(c) \subseteq S_p(c), \tag{6.18}$$

of cardinality

$$k_p(c) := |K_p(c)| \geq \log_2 \sqrt{N_p(c)},\tag{6.19}$$

such that all pairwise correlations of the random variables indexed by $K_p(c)$ are greater than $c^2/2$. Since the sequence $N_p(c) \to \infty$, by (6.19), we have $k_p(c) \to \infty$ as $p \to \infty$.

Therefore, we have identified a sequence of subsets $K_p(c) \subseteq \{1, \ldots, p\}$ with the following two properties:

1. $k_p(c) := |K_p(c)| \to \infty$, as $p \to \infty$, and
2. For all $i, j \in K_p(c)$, we have

$$\Sigma_p(i, j) > c^2/2.\tag{6.20}$$

Without loss of generality, we may assume $K_p(c) = \{1, \ldots, k_p(c)\} \subseteq \{1, \ldots, p\}$, upon re-labeling of the coordinates.

Now consider a Gaussian sequence $\epsilon^* = \{\epsilon^*(j), \ j = 1, 2, \ldots\}$, independent of $\mathcal{E}$, defined as follows:

$$\epsilon^*(j) := Z\left(c/\sqrt{2}\right) + Z(j)\sqrt{1 - c^2/2}, \quad j = 1, 2, \ldots,$$

where Z and $Z(j)$, $j = 1, 2, \ldots$ are independent standard normal random variables. Hence,

$$\mathrm{Var}(\epsilon^*(j)) = 1 = \mathrm{Var}(\epsilon_p(j)),\tag{6.21}$$

and

$$\mathrm{Cov}(\epsilon^*(i), \epsilon^*(j)) = \frac{c^2}{2} \leq \mathrm{Cov}(\epsilon_p(i), \epsilon_p(j)),\tag{6.22}$$

for all p, and all $i \neq j, i, j \in K_p(c)$. Thus, we have, as $p \to \infty$,

$$\frac{1}{u_{k_p(c)}} \max_{j \in K_p(c)} \epsilon^*(j) = \frac{c/\sqrt{2}}{u_{k_p(c)}} Z + \frac{\sqrt{1 - c^2/2}}{u_{k_p(c)}} \max_{j \in K_p(c)} Z(j) \overset{\mathbb{P}}{\to} \sqrt{1 - \frac{c^2}{2}},\tag{6.23}$$

where the convergence in probability follows from Proposition 2.2 part 2.

Relations (6.21) and (6.22), by Slepian's Lemma (recall Theorem 2.1), also imply,

$$\frac{1}{u_{k_p(c)}} \max_{j \in K_p(c)} \epsilon^*(j) \overset{d}{\geq} \frac{1}{u_{k_p(c)}} \max_{j \in K_p(c)} \epsilon_p(j).\tag{6.24}$$

Therefore, by (6.24) and (6.23), for all $\sqrt{1 - c^2/2} < \delta < 1$, we have

$$\mathbb{P}\left[\frac{1}{u_{k_p(c)}} \max_{j \in K_p(c)} \epsilon_p(j) < \delta\right] \to 1 \quad \text{as } p \to \infty.$$

This contradicts the definition of URS (with the particular choice of $S_p := K_p(c)$), and the proof of the "only if" part of Theorem 6.1 is complete.

6.3   UDD Implies URS (Proof of the 'If' Part of Theorem 6.1)

Recall that our objective is to show (4.21). We will do so in two stages; namely, we will prove that for all $\delta > 0$, we have

$$\mathbb{P}\left[\frac{M_{S_p}}{u_{|S_p|}} > 1 + \delta\right] \to 0, \tag{6.25}$$

and

$$\mathbb{P}\left[\frac{M_{S_p}}{u_{|S_p|}} < 1 - \delta\right] \to 0, \tag{6.26}$$

for any sequence of subsets S_p such that $|S_p| \to \infty$. Although the first step (6.25) was already shown in Proposition 2.2, regardless of the dependence structure, we provide in this section a more refined result. Specifically, the following result states that for the AGG model, the constant δ in Proposition 2.2 can be replaced by a vanishing sequence $c_p \to 0$.

Lemma 6.2 (Upper tails of AGG maxima) *Let $\mathcal{E}$ be an array with marginal distribution $F \in AGG(\nu)$, $\nu > 0$. If we pick*

$$c_p = \frac{u_{p \log p}}{u_p} - 1, \tag{6.27}$$

where $u_p = F^{\leftarrow}(1 - 1/p)$, then we have $c_p > 0$, $c_p \to 0$, and

$$\mathbb{P}\left[\frac{M_p}{u_p} - (1 + c_p) > 0\right] \to 0. \tag{6.28}$$

The proof can be found in Sect. 6.3.1 below.

Since Lemma 6.2 holds regardless of the dependence structure, the same conclusions hold if one replaces M_p by $M_{S_p} = \max_{j \in S_p} \epsilon(j)$ and p by $q = q(p) = |S_p|$, where S_p is any sequence of sets such that $q \equiv |S_p| \to \infty$. This entails (6.25).

On the other hand, the proof of (6.26) uses a more elaborate argument based on the Sudakov–Fernique bound. We proceed by first bounding the probability by an expectation. For all $\delta > 0$, we have

$$\mathbb{P}\left[\frac{M_{S_p}}{u_q} < 1 - \delta\right] = \mathbb{P}\left[-\left(\frac{M_{S_p}}{u_q} - (1 + c_q)\right) > \delta + c_q\right]$$

$$\leq \mathbb{P}\left[\left(\frac{M_{S_p}}{u_q} - (1 + c_q)\right)_- > \delta + c_q\right]$$

$$\leq \frac{1}{\delta + c_q}\mathbb{E}\left[\left(\frac{M_{S_p}}{u_q} - (1 + c_q)\right)_-\right], \qquad (6.29)$$

where $(x)_- := \max\{-x, 0\}$ and the last line follows from the Markov inequality. The next result shows that the upper bound in (6.29) vanishes.

Lemma 6.3 *Let $\mathcal{E}$ be a Gaussian UDD array and $S_p \subseteq \{1, \dots, p\}$ be an arbitrary sequence of sets such that $q = q(p) = |S_p| \to \infty$. Then, for $M_{S_p} := \max_{j \in S_p} \epsilon_p(j)$ and c_q as in (6.27), we have*

$$\mathbb{E}\left[\left(\frac{M_{S_p}}{u_q} - (1 + c_q)\right)_-\right] \to 0, \qquad as\ p \to \infty. \qquad (6.30)$$

The proof of the lemma is given in Sect. 6.3.2 below.

Going back to the proof of Theorem 6.1, we observe that Relations (6.29) and (6.30) imply (6.26), which completes the proof of the 'if' part of Theorem 6.1. $\square$

Remark 6.3 Only the Sudakov–Fernique minorization argument used in the proof of Lemma 6.3, relies on the Gaussian assumption. We expect the techniques and results here to be useful in extending Theorem 6.1 to more general class of distributions, say, the AGG model.

6.3.1 Bounding the Upper Tails of AGG Maxima

Proof (*Lemma 6.2*) Recall by (2.33) that

$$u_q \sim (v \log q)^{1/v}, \quad q \to \infty,$$

so that

$$c_p = \frac{u_{p \log p}}{u_p} - 1 = \left(\frac{\log p + \log \log p}{\log p}\right)^{1/v}(1 + o(1)) - 1 \to 0 \quad as\ p \to \infty.$$

$$\qquad (6.31)$$

By the union bound, we have

$$\mathbb{P}\left[\frac{M_p}{u_p} > 1 + c_p\right] \leq \sum_{j=1}^{p} \mathbb{P}\left[\frac{\epsilon_p(j)}{u_p} > 1 + c_p\right] = p\overline{F}\left(u_{p\log p}\right) \qquad (6.32)$$

$$= p\overline{F}\left(F^{\leftarrow}\left(1 - \frac{1}{p\log p}\right)\right) \leq \frac{1}{\log p} \to 0,$$

where the last inequality follows from the fact that $F\left(F^{\leftarrow}(u)\right) \geq u$ for all $u \in [0, 1]$. $\square$

In addition to Lemma 6.2, which says the upper tail vanishes in probability, we will also prepare a result which states that the upper tail vanishes in expectation.

Lemma 6.4 *Let M_p and c_p be as in Lemma 6.2, and denote*

$$\xi_p := \frac{M_p}{(1 + c_p)u_p}.$$

Then there exist $p_0, t_0 > 0$, and an absolute constant $C > 0$ such that

$$\mathbb{P}\left[\xi_p > t\right] \leq \exp\{-Ct^\nu\}, \quad \text{for all} \quad p > p_0, \, t > t_0. \qquad (6.33)$$

In particular, the set of random variables $\{(\xi_p)_+ , p \in \mathbb{N}\}$ is uniformly integrable.

Proof *(Lemma 6.4)* Recall that $(1 + c_p)u_p = u_{p\log p}$, and by applying the union bound as in (6.32), we have

$$\log \mathbb{P}\left[\xi_p > t\right] \leq \log p + \log \overline{F}\left(u_{p\log p}t\right)$$

$$\leq \log p - \frac{1}{\nu}\left(u_{p\log p}t\right)^\nu (1 - \delta), \qquad (6.34)$$

for $t > t_0(\delta) > 0$, where $\delta \in (0, 1)$ is an arbitrarily small number fixed in advance. This follows from the assumption that $F \in \text{AGG}(\nu)$ and Definition 2.6 of the AGG distribution. Using in (6.34), the explicit expressions for the quantiles in (2.33), we obtain

$$\log \mathbb{P}\left[\xi_p > t\right] \leq \log p - \underbrace{(1 + o(1))\,(1 - \delta)t^\nu}_{\text{greater than 1 for large } t}\log p - t^\nu \underbrace{\log\log p\,(1 + o(1))\,(1 - \delta)}_{\text{greater than } C \text{ for large } p}.$$

$$(6.35)$$

For large t, we have $(1 + o(1))\,(1 - \delta)t^\nu > 1$ so that the sum of the first two terms on the right-hand side of (6.35) is negative. Also, for p larger than some constant $p_0(\delta)$, we have $\log\log p\,(1 + o(1))\,(1 - \delta) > C$ for some constant C that does not depend on p. Therefore (6.33) holds for $t > t_0(\delta)$ and $p > p_0(\delta)$, and the proof is complete. $\square$

Corollary 6.3 *The upper tails of AGG maxima vanish in expectation, i.e.,*

$$\mathbb{E}\left[\left(\frac{M_p}{u_p} - (1 + c_p)\right)_+\right] \to 0 \quad as \ p \to \infty, \tag{6.36}$$

where $(a)_+ := \max\{a, 0\}$.

Proof *(Corollary* 6.3*)* Since $c_p \geq 0$ is a sequence converging to 0, we have $c_p < 1$ for $p \geq p_0$. Hence, for any $t > 0$, we have

$$\mathbb{P}\left[\left(\frac{M_p}{u_p} - (1 + c_p)\right)_+ > t\right] = \mathbb{P}\left[(1 + c_p)\left(\xi_p - 1\right)_+ > t\right]$$

$$\leq \mathbb{P}\left[\left(\xi_p - 1\right)_+ > t/2\right] \leq \mathbb{P}\left[\xi_p > t/2\right]. \tag{6.37}$$

By Lemma 6.4, $\{(\xi_p)_+\}$ is u.i., therefore by Relation (6.37), $\{(M_p/u_p - (1 + c_p))_+,$ $p \in \mathbb{N}\}$ is u.i. as well. Since by Lemma 6.2, $(M_p/u_p - (1 + c_p))_+ \to 0$ in probability, Relation (6.36) follows from the established uniform integrability (see, e.g., Theorem 6.6.1 in Resnick 2014). □

6.3.2 Bounding the Lower Tails of Gaussian Maxima

The main goal of this section is to establish the following result.

Proposition 6.3 *For every UDD Gaussian array $\mathcal{E}$, and any sequence of subsets $S_p \subseteq \{1, \ldots, p\}$ such that $q = q(p) = |S_p| \to \infty$, we have*

$$\liminf_{p \to \infty} \mathbb{E}\left[\frac{M_{S_p}}{u_q}\right] \geq 1, \tag{6.38}$$

where $M_S = \max_{j \in S} \epsilon(j)$.

We will first show that Lemma 6.3, which is the key to the proof of the 'if' part of Theorem 6.1, follows immediately from this proposition.

Proof *(Lemma* 6.3*)* We start with the identity

$$\mathbb{E}\left[\frac{M_{S_p}}{u_q} - (1 + c_q)\right] = \mathbb{E}\left[\left(\frac{M_{S_p}}{u_q} - (1 + c_q)\right)_+\right] - \mathbb{E}\left[\left(\frac{M_{S_p}}{u_q} - (1 + c_q)\right)_-\right].$$

By re-arranging terms and taking limsup/liminf, we obtain

$$0 \leq \limsup_{p \to \infty} \mathbb{E}\left[\left(\frac{M_{S_p}}{u_q} - (1 + c_q)\right)_{-}\right]$$

$$\leq \limsup_{p \to \infty} \mathbb{E}\left[\left(\frac{M_{S_p}}{u_q} - (1 + c_q)\right)_{+}\right] - \liminf_{p \to \infty} \mathbb{E}\left[\frac{M_{S_p}}{u_q} - (1 + c_q)\right] \quad (6.39)$$

$$= -\liminf_{p \to \infty} \mathbb{E}\left[\frac{M_{S_p}}{u_q} - (1 + c_q)\right], \qquad (6.40)$$

where the last equality follows from the fact that the lim-sup in (6.39) vanishes by Corollary 6.3. On the other hand, since $c_q \to 0$, we have

$$\liminf_{p \to \infty} \mathbb{E}\left[\frac{M_{S_p}}{u_q} - (1 + c_q)\right] = \liminf_{p \to \infty} \mathbb{E}\left[\frac{M_{S_p}}{u_q} - 1\right] \geq 0,$$

where the last inequality follows from Proposition 6.3. This shows that the right-hand side of (6.40) is non-positive and hence (6.30) holds. $\qquad \square$

The following interesting fact about the relationship between the upper quantiles and the expectation of iid maxima will be needed for the proof of Proposition 6.3.

Lemma 6.5 *Let $(X_i)_{i=1}^{p}$ be p iid random variables with distribution F such that $\mathbb{E}[(X_i)_{-}]$ exists, i.e.,*

$$\mathbb{E}[\max\{-X_i, 0\}] < \infty.$$

Let $M_p = \max_{i=1,\dots,p} X_i$. Assume that F has a density f, which is eventually decreasing. More precisely, we suppose there exists a C_0 such that $0 < F(C_0) < 1$, and $f(x_1) \geq f(x_2)$ whenever $C_0 < x_1 \leq x_2$. Under these assumptions, we have,

$$\liminf_{p \to \infty} \frac{\mathbb{E}M_p}{u_{p+1}} \geq 1,$$

where $u_{p+1} = F^{\leftarrow}(1 - 1/(p + 1))$.

Proof The idea comes from an argument in the monograph of Boucheron et al. (2013). Write

$$X_i = F^{\leftarrow}(U_i)$$

where U_i are iid uniform random variables on $(0, 1)$. Denote M_p^U as the maximum of the U_i's, we have $\mathbb{E}M_p = \mathbb{E}\left[F^{\leftarrow}(M_p^U)\right]$, and by conditioning, we obtain

$$\mathbb{E}M_p = \mathbb{E}\left[F^{\leftarrow}(M_p^U) \mid M_p^U \geq F(C_0)\right] \mathbb{P}\left[M_p^U \geq F(C_0)\right] + \\ + \mathbb{E}\left[F^{\leftarrow}(M_p^U) \mid M_p^U < F(C_0)\right] \mathbb{P}\left[M_p^U < F(C_0)\right]. \qquad (6.41)$$

Focus on the first term in the summation. Since f is decreasing beyond C_0, F is concave on (C_0, ∞), and $F^\leftarrow$ is convex on $(F(C_0), 1)$. By Jensen's inequality, we have

$$\mathbb{E}\left[F^\leftarrow(M_p^U) \mid M_p^U \geq F(C_0)\right] \geq F^\leftarrow\left(\mathbb{E}[M_p^U \mid M_p^U \geq F(C_0)]\right).$$

Using the fact that M_p^U is the maximum of iid Uniform$(0, 1)$ random variables, with a direct calculation one can show that

$$F^\leftarrow\left(\mathbb{E}[M_p^U \mid M_p^U \geq F(C_0)]\right) = F^\leftarrow\left(\left(1 - \frac{1}{p+1}\right)\left(\frac{1 - F(C_0)^{p+1}}{1 - F(C_0)^p}\right)\right),$$

and hence

$$\mathbb{E}\left[F^\leftarrow(M_p^U) \mid M_p^U \geq F(C_0)\right] \geq F^\leftarrow\left(\left(1 - \frac{1}{p+1}\right)\left(\frac{1 - F(C_0)^{p+1}}{1 - F(C_0)^p}\right)\right)$$

$$\geq F^\leftarrow\left(1 - \frac{1}{p+1}\right) = u_{p+1}. \tag{6.42}$$

Now, focus on the second term in (6.41). Since $\mathbb{P}[M_p^U \leq m \mid M_p^U < F(C_0)] = (m/F(C_0))^p \leq m/F(C_0)$ for $m \leq F(C_0)$, we have

$$\left(M_p^U \mid M_p^U < F(C_0)\right) \overset{\mathrm{d}}{\geq} \left(U_1 \mid U_1 < F(C_0)\right),$$

where and the latter is the uniform distribution on $(0, F(C_0))$. Therefore, for the second term of the sum in (6.41), by the monotonicity of $F^\leftarrow$, we obtain

$$\mathbb{E}\left[F^\leftarrow(M_p^U) \mid M_p^U < F(C_0)\right] \geq \mathbb{E}\left[F^\leftarrow(U_1) \mid U_1 < F(C_0)\right]$$

$$= \mathbb{E}\left[X_1 \mid X_1 < C_0\right]. \tag{6.43}$$

Finally, since $\mathbb{P}\left[M_p^U < F(C_0)\right] = F(C_0)^p = 1 - \mathbb{P}\left[M_p^U \geq F(C_0)\right]$, by (6.42) and (6.43), we have

$$\frac{\mathbb{E}M_p}{u_{p+1}} \geq \left(1 - F(C_0)^p\right) + \frac{\mathbb{E}\left[X_1 \mid X_1 < C_0\right]}{u_{p+1}} F(C_0)^p.$$

The conclusion follows since the right-hand side of the last inequality converges to 1.   □

We are now ready to prove Proposition 6.3. This is where the UDD dependence assumption is used.

Proof (*Proposition* 6.3) Recall that $\mathcal{E} = \{\epsilon_p(i),\ i = 1, \cdots, p,\ p \in \mathbb{N}\}$ is a Gaussian array with standardized marginals. Define the canonical (pseudo) metric on $S_p \subset \{1, \cdots, p\}$,

$$d(i, j) = \sqrt{\mathbb{E}\left[(\epsilon(i) - \epsilon(j))^2\right]}.$$

It can be easily checked that the canonical metric takes values between 0 and 2. For an arbitrary $\delta \in (0, 1)$, take $\gamma = \sqrt{2(1 - \delta)}$, and let $\mathcal{N}$ be a γ-packing of S_p. That is, let $\mathcal{N}$ be a subset of S_p, such that for any $i, j \in \mathcal{N}, i \neq j$, we have $d(i, j) \geq \gamma$, i.e.,

$$d(i, j) = \sqrt{2\left(1 - \Sigma_p(i, j)\right)} \geq \gamma = \sqrt{2(1 - \delta)}, \qquad (6.44)$$

or equivalently, $\Sigma_p(i, j) \leq \delta$. We claim that we can find a γ-packing $\mathcal{N}$ whose number of elements is at least

$$|\mathcal{N}| \geq q/N(\delta). \qquad (6.45)$$

Indeed, $\mathcal{N}$ can be constructed iteratively as follows:

1: Set $S_p^{(1)} := S_p$ and $\mathcal{N} := \{j_1\}$, where $j_1 \in S_p^{(1)}$ is an arbitrary element. Set $k := 1$.

2: Set $S_p^{(k+1)} := S_p^{(k)} \setminus B_\gamma(j_k)$, where

$$B_\gamma(j_k) := \{i \in S_p : \ d(i, j_k) < \gamma \equiv \sqrt{2(1 - \delta)}\}.$$

3: If $S_p^{(k)} \neq \emptyset$, pick an arbitrary $j_{k+1} \in S_p^{(k)}$, set $\mathcal{N} := \mathcal{N} \cup \{j_{k+1}\}$, and $k := k + 1$, go to step 2; otherwise, stop.

By the definition of UDD (see Definition 6.1), there are at most $N(\delta)$ coordinates whose covariance with $\epsilon(j)$ exceed δ. Therefore at each iteration, $\left|B_\gamma(j_k)\right| \leq N(\delta)$, and hence

$$\left|S_p^{(k+1)}\right| \geq \left|S_p^{(k)}\right| - \left|B_\gamma(j_k)\right| \geq q - kN(\delta).$$

The construction can continue for at least $q/N(\delta)$ iterations, and we have $|\mathcal{N}| \geq \lfloor q/N(\delta) \rfloor$ as desired.

Now we define on this γ-packing $\mathcal{N}$ an independent Gaussian process $(\eta(j))_{j \in \mathcal{N}}$,

$$\eta(j) = \frac{\gamma}{\sqrt{2}} Z(j) \quad j \in \mathcal{N},$$

where $Z(j)$'s are iid standard normal random variables. Observe that by the definition of γ-packing in (6.44), the increments of the new process are smaller than those of the original process in the following sense,

$$\mathbb{E}\left[(\eta(i) - \eta(j))^2\right] = \gamma^2 \leq d^2(i, j) = \mathbb{E}\left[(\epsilon(i) - \epsilon(j))^2\right]$$

for all $i \neq j, i, j \in \mathcal{N}$. Applying the Sudakov–Fernique inequality (see Theorem 2.2) to $(\eta(j))_{j \in \mathcal{N}}$ and $(\epsilon(j))_{j \in \mathcal{N}}$, we have

$$\mathbb{E}\left[\max_{j \in \mathcal{N}} \eta(j)\right] \leq \mathbb{E}\left[\max_{j \in \mathcal{N}} \epsilon(j)\right] \leq \mathbb{E}\left[\max_{j \in S_p} \epsilon(j)\right]. \tag{6.46}$$

Since the $(\eta(j))_{j \in \mathcal{N}}$ are independent Gaussians, Lemma 6.5 yields the lower bound,

$$\liminf_{p \to \infty} \mathbb{E}\left[\frac{\max_{j \in \mathcal{N}} \eta(j)}{u_{|\mathcal{N}|}}\right] \geq \frac{\gamma}{\sqrt{2}} = \sqrt{1 - \delta}. \tag{6.47}$$

Using (6.45) and the expressions (2.33) for the quantiles of AGG models (with $\nu = 2$ here), we have

$$\frac{u_{|\mathcal{N}|}}{u_q} \geq \left(\frac{\log q - \log N(\delta)}{\log q}\right)^{1/2} (1 + o(1)) \to 1, \tag{6.48}$$

since $N(\delta)$ does not depend on $q = q(p) \to \infty$.

By combining (6.46), (6.47) and (6.48), we conclude that

$$\liminf_{p \to \infty} \mathbb{E}\left[\frac{\max_{j \in S_p} \epsilon(j)}{u_q}\right] \geq \liminf_{p \to \infty} \mathbb{E}\left[\frac{\max_{j \in \mathcal{N}} \eta(j)}{u_q}\right] \qquad \text{by (6.46)}$$

$$\geq \liminf_{p \to \infty} \mathbb{E}\left[\frac{\max_{j \in \mathcal{N}} \eta(j)}{u_{|\mathcal{N}|}}\right] \qquad \text{by (6.48)}$$

$$\geq \sqrt{1 - \delta}. \qquad \text{by (6.47)}$$

Since $\delta > 0$ is arbitrary, (6.38) follows as desired. $\qquad\square$

6.4 Numerical Illustrations of Exact Support Recovery Under Dependence

The characterization of URS with the UDD condition allows us to simulate Gaussian errors and illustrate the effect of dependence on the phase transition behavior in finite dimensions. We shall compare the performance of the Bonferroni's procedure, which is agnostic to both sparsity and signal size, with the oracle procedure which picks the top-s observations.

The first set of experiments explores short-range dependent errors from an autoregressive (AR) models.

- AR(1) Gaussian errors with parameter $\rho = -0.5$, $\rho = 0.5$, and $\rho = 0.9$, where the autocovariance functions decay exponentially, $\rho_k = \rho^k$.

We apply both the sparsity- and signal-size agnostic Bonferroni's procedure, i.e., $\widehat{S} = \{i : x(i) > \sqrt{2 \log p}\}$, as well as the oracle procedure $\widehat{S}^* = \{i : x(i) \geq x_{[s]}\}$, $s = |S|$, to all settings. Results of the numerical experiments for the AR models are shown in Fig. 6.1.

For dependent errors the oracle procedures is able to recover support of signals with higher probability than the Bonferroni procedures in finite dimensions; compare left and right columns of Fig. 6.1. For short range dependent observations, however, there is not a pronounced difference. The results of the experiments are very similar to that of the independent Gaussian case.

The second set of experiments explores exact support recovery in additive error models in the cases of long-range dependent but UDD errors, as well as non-UDD errors. In particular we simulate

- Fractional Gaussian noise (fGn) with Hurst parameter $H = 0.75$ and $H = 0.9$. The autocovariance functions are

$$\rho_k \sim 0.75 k^{-0.6} \quad \text{and} \quad \rho_k \sim 1.44 k^{-0.2},$$

as $k \to \infty$. Both fGn models represent the regime of long-range dependence, where covariances decay very slowly to zero, so that $\sum |\rho_k| = \infty$; see, e.g., Taqqu (2003). Observe that every stationary Gaussian process with vanishing autocovariance gives rise to an UDD array as concluded in Corollary 6.1.
- The non-UDD Gaussian errors described in Example 6.1.

We again apply both the sparsity-and-signal-size-agnostic Bonferroni's procedure, i.e., $\widehat{S} = \{i : x(i) > \sqrt{2 \log p}\}$, as well as the oracle procedure $\widehat{S}^* = \{i : x(i) \geq x_{[s]}\}$, $s = |S|$, to all settings. Results of the numerical experiments for the fGn and non-UDD models are shown in Fig. 6.2.

Notice that the oracle procedure sets its thresholds more aggressively (at roughly $\sqrt{2 \log s}$) than the Bonferroni procedure (at $\sqrt{2 \log p}$). Although this difference vanishes as $p \to \infty$, in finite dimensions ($p = 10\,000$) the advantage can be felt. Indeed, in all our experiments the oracle procedure is able to recover support of signals with higher probability than the Bonferroni procedures; compare left and right columns of Fig. 6.2. Notice also that there is an increase in probability of recovery near $\beta = 0$ for oracle procedures. This is an artifact in finite dimensions due to the fact that $s = \lfloor p^{1-\beta} \rfloor < p/2$, and there are more signals than nulls. The oracle procedures is able to adjust to this reversal by lowering its threshold accordingly.

For UDD errors, Theorem 4.2 predicts that exact recovery of the support is impossible when signal sizes are below the boundary (4.5), even with oracle procedures. However, the rate of this convergence (i.e., $\mathbb{P}[\widehat{S}^* = S] \to 0$ or 1) can be very slow when the errors are heavily dependent, even though all AR and fGn models demonstrate qualitatively the same behavior in line with the predicted boundary (4.5). In finite dimensions ($p = 10\,000$), as dependence in the errors increases (fGN(H = 0.75) to fGN(H = 0.9)), the oracle procedure becomes more powerful at recovering signal support with high probability for weaker signals.

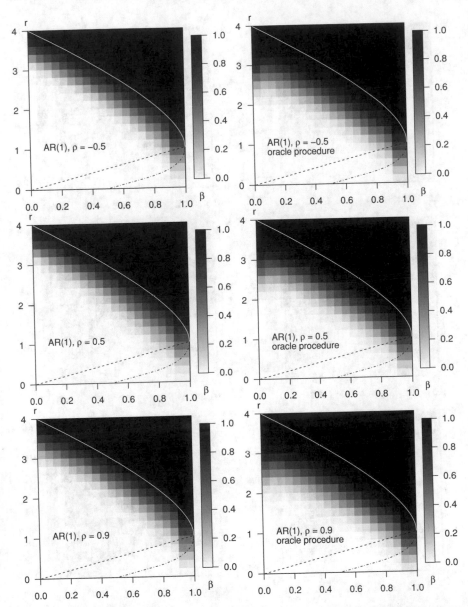

Fig. 6.1 The empirical probability of exact support recovery from numerical experiments, as a function of sparsity level β and signal sizes r. Darker colors indicate higher probability of exact support recovery. Three AR(1) models with autocorrelation functions $(-0.5)^k$ (upper), 0.5^k (middle), and 0.9^k (lower) are simulated. The experiments were repeated 1000 times for each sparsity-signal size combination. In finite dimensions ($p = 10000$), the Bonferroni procedures (left) suffers small loss of power compared to the oracle procedures (right). A phase transition in agreement with the predicted boundary (4.5) can be seen in the AR models. The boundaries (solid, dashed, and dash-dotted lines) are as in Fig. 4.1

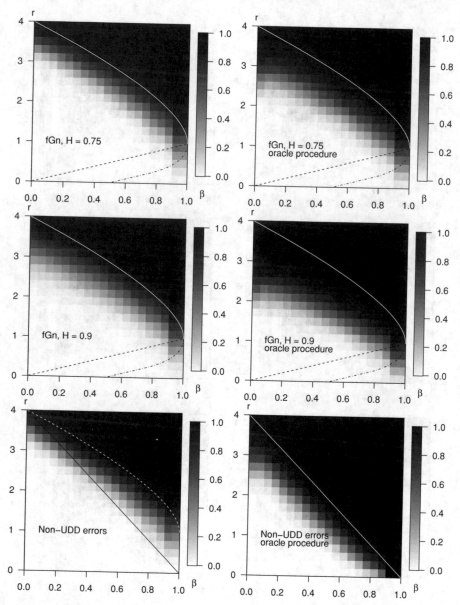

Fig. 6.2 The empirical probability of exact support recovery from numerical experiments, as a function of sparsity level β and signal sizes r. Darker colors indicate higher probability of exact support recovery. Two fGn models with Hurst parameter $H = 0.75$ (upper), $H = 0.9$ (middle), and the non-UDD errors in Example 6.1 (lower) are simulated. The experiments were repeated 1000 times for each sparsity-signal size combination. In finite dimensions ($p = 10000$), the oracle procedures (right) is able to recover support for weaker signals than the Bonferroni procedures (left) when errors are heavily dependent, although they have the same phase transition limit. The non-UDD errors demonstrate qualitatively different behavior, enabling support recovery for strictly weaker signals. The boundaries (solid, dashed, and dash-dotted lines) are as in Fig. 4.1. In the non-UDD example, dashed lines represent the limit attained by Bonferroni's procedures. See text for additional comments

On the other hand, as demonstrated in Example 6.1, non-UDD errors yield qualitatively different behavior; exact support recovery is possible for signal sizes strictly weaker than that in the UDD case. Lower-right panel of Fig. 6.2 demonstrates in this example that the signal support can be recovered as long as the signal sizes are larger than $4(1 - \beta)$.

Chapter 7
Fundamental Statistical Limits
in Genome-Wide Association Studies

The process of scientific discovery, as explained by Richard Feynman, usually starts with guesses. The consequences of such guesses are then computed and compared with experimental results. If the predictions disagree with the experiment then our guesses are wrong. "That is all there is to it" (Feynman 2017).

In the previous chapters, we delved deep into the theoretical underpinnings of the phase-transition phenomena in high-dimensional multiple testing problems. The results are interesting in their own right. However, we have not discovered any scientific law in the spirit of Feynman, but merely worked out mathematical consequences of our postulated models. In this chapter, our goal is to relate these predictions to real experimental data from the field of genetics, where large-scale simultaneous hypotheses testing problems often arise. From such comparisons, we will demonstrate that the phase transition laws are indeed reasonable predictions of some curious phenomena in that field. The accuracy of our predictions will lend credibility to the application of these "laws of large dimensions" in actual applications.

In our case, the experimental data used as the measuring stick come from genome-wide association studies (GWAS), introduced in Sect. 1.2. Recall that in GWAS, a large number of marginal association tests are conducted simultaneously, resulting in statistics that can be approximated by

$$x(i) \sim \chi_\nu^2(\lambda(i)), \quad i = 1, \ldots, p, \tag{7.1}$$

where $\chi_\nu^2(\lambda(i))$ is a chi-square distributed random variable with $\nu > 0$ degrees of freedom[1] and non-centrality parameter $\lambda(i)$.

We establish our theoretical predictions in two steps. In Sect. 7.1 below, we shall first establish the phase transitions of the model (7.1). In parallel to results in Chap. 3,

---

[1] The parameter ν here should not be confused with the shape parameter of the AGG(ν) distributions, which will not appear in this chapter.

© The Author(s), under exclusive license to Springer Nature Switzerland AG 2021
Z. Gao and S. Stoev, *Concentration of Maxima and Fundamental
Limits in High-Dimensional Testing and Inference*,
SpringerBriefs in Probability and Mathematical Statistics,
https://doi.org/10.1007/978-3-030-80964-5_7

we show that several commonly used family-wise error rate-control procedures—including Bonferroni's procedure—are asymptotically optimal for the exact, and the exact–approximate support recovery problems (recall Definition 2.5) in idealized chi-square models with independent components. Analogously, the BH procedure is asymptotically optimal for the approximate, and the approximate–exact support recovery problems. Under appropriate parametrization of the signal sizes and sparsity, we establish the phase transitions of support recovery problems in the chi-square model. Remarkably, the degree-of-freedom parameter does not affect the asymptotic boundaries in any of the four support recovery problems.

In the second step, we translate the canonical signal size and sparsity parametrizations into the vernacular of statistical geneticists in Sect. 7.2. We do so by characterizing the relationship between the signal size λ and the marginal frequencies, odds ratio, and sample sizes for association tests on 2-by-2 contingency tables. This is important because the latter parameters are often estimated and reported in GWAS, while we have never seen the elusive signal size parameter λ reported. As a bonus, we point out the implications of this relationship on statistically optimal study designs for association studies in Sect. 7.3: perhaps surprisingly, balanced designs with equal number of cases and controls are often statistically inefficient.

Armed with the results on phase transitions in the chi-square model, and a translation from the language of high-dimensional statistics to the patois of association screening studies, we finally present in Sect. 7.4 the consequences of the phase transitions in GWAS, and compare against real experimental data to evaluate the success of our predictions.

The phase transitions in the chi-square models are demonstrated with numerical simulations in Sect. 7.5. The proofs, which are closely resemble those of the results in Chap. 3, are collected in Appendix A.

7.1 Support Recovery Problems in Chi-Squared Models

Similar to the analysis of additive error models in Chap. 3, we will work with triangular arrays of chi-square models (7.1) indexed by p. We adopt the same parametrization for the sparsity of the non-centrality parameter vectors $\lambda = \lambda_p$,

$$|S_p| = \lfloor p^{1-\beta} \rfloor, \quad \beta \in (0, 1] \tag{7.2}$$

where $S_p := \{i : \lambda(i) > 0\}$ is the *signal support* set and β parametrizes the problem *sparsity*. More general parameterizations of the support size are possible as in (4.3). Here, however, we drop the slowly varying term $\ell(\cdot)$ for simplicity. The closer β is to 1, the sparser the support S_p; conversely, when β is close to 0, the support is dense with many non-null signals.

We parametrize the range of the non-zero and perhaps unequal signals in the chi-square model with

$$\underline{\Delta} = 2\underline{r} \log p \leq \lambda(i) \leq \overline{\Delta} = 2\overline{r} \log p, \quad \text{for all } i \in S_p, \tag{7.3}$$

for some constants $0 < \underline{r} \leq \overline{r} \leq +\infty$.

7.1.1 The Exact Support Recovery Problem

The first main result characterizes the phase transition phenomenon in the exact support recovery problem under the chi-square model. It parallels Theorem 3.2.

Theorem 7.1 *Consider the high-dimensional chi-squared model* (7.1) *with signal sparsity and size as described in* (7.2) *and* (7.3). *The function*

$$f_E(\beta) = \left(1 + \sqrt{1 - \beta}\right)^2 \tag{7.4}$$

characterizes the phase transition of exact support recovery problem. Namely, the following two results hold.

(i) *If* $\underline{r} > f_E(\beta)$, *then Bonferroni's, Sidák's, Holm's, and Hochberg's procedures with slowly vanishing (see Definition 3.1) nominal FWER levels all achieve asymptotically exact support recovery in the sense of* (2.25).
(ii) *Conversely, if* $\overline{r} < f_E(\beta)$, *then for any thresholding procedure* $\widehat{S}_p$, *we have* $\mathbb{P}[\widehat{S}_p = S_p] \to 0$. *Therefore, in view of Lemma 2.1, exact support recovery asymptotically fails for all thresholding procedures in the sense of* (2.26).

The procedures listed in Theorem 7.1 were reviewed in Sect. 2.2. The proof of the theorem can be found in Sect. A.2.

It is evident that the exact support recovery boundary (7.4) coincides with that of the Gaussian additive error models (1.1) in Chap. 3. Implications of these results will be discussed in Sect. 7.1.5 below.

Remark 7.1 Theorem 7.1 predicts that the asymptotic boundaries are the same for all values of the degrees of freedom parameter ν. In simulations (Sect. 7.5), we find this asymptotic prediction to be quite accurate for $\nu \leq 3$ even in moderate dimensions ($p = 100$). For $\nu > 3$, the phase transitions take place somewhat above the boundary g. The behavior is qualitatively similar in the other three phase transitions (see Theorems 7.2, 7.3, and 7.4 below).

7.1.2　The Exact–Approximate Support Recovery Problem

The next theorem describes the phase transition in the exact–approximate support recovery problem. Recall also Theorem 3.4.

Theorem 7.2 *In the context of Theorem 7.1, the function*

$$f_{\mathrm{EA}}(\beta) = 1 \tag{7.5}$$

characterizes the phase transition of exact–approximate support recovery problem. Namely, the following two results hold.

(i) *If $\underline{r} > f_{\mathrm{EA}}(\beta)$, then the procedures listed in Theorem 7.1 with slowly vanishing nominal FWER levels achieve asymptotically exact–approximate support recovery in the sense of (2.25).*

(ii) *Conversely, if $\overline{r} < f_{\mathrm{EA}}(\beta)$, then for any thresholding procedure $\widehat{S}_p$, the exact–approximate support recovery fails in the sense of (2.26).*

Theorem 7.2 is proved in Sect. A.4.

7.1.3　The Approximate Support Recovery Problem

Our third asymptotic result characterizes the phase-transition phenomenon in the approximate support recovery problem in the chi-square model. It closely parallels Theorem 3.3 for the additive errors model.

Theorem 7.3 *Consider the high-dimensional chi-squared model (7.1) with signal sparsity and size as described in (7.2) and (7.3). The function*

$$f_{\mathrm{A}}(\beta) = \beta \tag{7.6}$$

characterizes the phase transition of approximate support recovery problem. Specifically, the following two results hold.

(i) *If $\underline{r} > f_{\mathrm{A}}(\beta)$, then the BH procedure $\widehat{S}_p$ (defined in Sect. 2.2) with slowly vanishing (see Definition 3.1) nominal FDR levels achieves asymptotically approximate support recovery in the sense of (2.25).*

(ii) *Conversely, if $\overline{r} < f_{\mathrm{A}}(\beta)$, then approximate support recovery asymptotically fails in the sense of (2.26) for all thresholding procedures.*

Theorem 7.3 is proved in Sect. A.4.

7.1.4 The Approximate–Exact Support Recovery Problem

A counterpart of Theorem 3.5 also holds in the chi-square models.

Theorem 7.4 *In the context of Theorem 7.3, the function*

$$f_{\mathrm{AE}}(\beta) = \left(\sqrt{\beta} + \sqrt{1-\beta}\right)^2 \tag{7.7}$$

characterizes the phase transition of approximate–exact support recovery problem. Namely, the following two results hold.

(i) *If $\underline{r} > f_{\mathrm{AE}}(\beta)$, then the Benjamini–Hochberg procedure with slowly vanishing nominal FDR levels achieves asymptotically approximate–exact support recovery in the sense of (2.25).*

(ii) *Conversely, if $\overline{r} < f_{\mathrm{AE}}(\beta)$, then for any thresholding procedure $\widehat{S}_p$, the approximate–exact support recovery fails in the sense of (2.26).*

Theorem 7.4 is proved in Sect. A.2.

Notice that all phase-transitions boundaries are identical to those in the Gaussian additive error model (1.1) under one-sided alternative. We refer readers to Fig. 3.2 in Sect. 3.2 for a visualization of the results in Theorems 7.1 through 7.4.

All four phase transitions results in Theorems 7.1 through 7.4 focus only on the idealized models (7.1) where the statistics are *independent*. Support recovery problems under dependent observations remain to be explored. Recall in Chap. 4 we showed that the boundary for the exact support recovery problem in the additive error model (1.1) continues to hold even under *severe dependence* and general distributional assumptions. We conjecture that the concentration of maxima phenomenon, which is at the heart of the results in Chap. 4, will play a role and all of the above phase-transition results will continue to hold, under broad dependence conditions in the chi-square models. As an example, in the GWAS application, dependence among the genetic markers at different locations (known as linkage disequilibrium) decay as a function of their physical distances on the genome (Bush and Moore 2012), resulting in locally dependent test statistics. It would be of great interest to extend the current theory to cover important dependence structures that arise in such applications.

7.1.5 Comparison of One- Versus Two-Sided Alternatives in Additive Error Models

As alluded to in Sect. 1.2 in the introduction, we draw explicit comparisons between the one-sided and two-sided alternatives in Gaussian additive error models (1.1).

The exact support recovery problem in the dependent Gaussian additive error model (1.1) was studied in Chap. 3, with parametrization of sparsity identical to that in (7.2), whereas the range of the non-zero (and perhaps unequal) mean shifts $\mu(i)$ was parametrized as

$$\underline{\Delta} = \sqrt{2\underline{r} \log p} \leq \mu(i) \leq \overline{\Delta} = \sqrt{2\overline{r} \log p}, \quad \text{for all } i \in S_p,$$

for some constants $0 < \underline{r} \leq \overline{r} \leq +\infty$. Under this one-sided alternative, a phase transition in the r-β plane was described, where the boundary was found to be identical to (7.4) in Theorem 7.1 for the chi-square models (7.1).

As discussed in Sect. 1.2, support recovery problems in the chi-square model with $\nu = 1$ correspond to the support recovery problems in the additive model under two-sided alternatives. This implies that the asymptotic signal size requirements are identical between the two-sided alternative and its one-sided counterpart, in order to achieve exact support recovery. As we shall see in numerical experiments (in Sect. 7.5 below), the difference is not very pronounced even in moderate dimensions, and vanishes as $p \to \infty$, in accordance with Theorem 7.1.

Comparisons can also be drawn in the approximate, the approximate–exact, and the exact–approximate support recovery problems between the two types of alternatives.

Specifically, the approximate support recovery problem in the Gaussian additive error model (1.1) under one-sided alternatives exhibits a phase transition phenomenon characterized by a boundary that coincides with (7.6) in Theorem 7.3. Similar to the exact support recovery problem, this indicates vanishing difference in the difficulties of the two types alternatives in approximate support recovery problems.

Comparing Theorems 7.2 and 3.4 as well as Theorems 7.4 and 3.5, we see that the phase transition boundaries under the two types of alternatives are also identical in the exact–approximate and approximate–exact support recovery problems.

To complete the comparisons, we point out that the phase-transition boundaries for the sparse signal detection problem in the two types of alternatives are both identical to (3.4). This was analyzed in Donoho and Jin (2004).

Therefore, all phase-transition boundaries coincide with those in the additive error models obtained in Chap. 3 under their respective parametrizations. This indicates vanishing differences between the difficulties of the one-sided and two-sided alternatives in the Gaussian additive error model (1.1). The additional uncertainty in the two-sided alternatives does not call for larger signal sizes in these problems, asymptotically.

7.2 Odds Ratios and Statistical Power

We return to the application of association screenings for categorical variables, and put the results in the previous section to use. In particular, we focus on the exact–approximate support recovery problem, and demonstrate the consequences of its phase transition (Theorem 7.2) in genetic association studies.

Table 7.1 Probabilities of the multinomial distribution in a genetic association test. (Compare and contrast with Table 1.1. We have $\mathbb{E}[O_{ij}] = n\mu_{ij}$, $i, j = 1, 2$, where $n = \sum_{i,j} O_{ij}$.)

Probabilities	Genotype		Total by phenotype
	Variant 1	Variant 2	
Cases	μ_{11}	μ_{12}	ϕ_1
Controls	μ_{21}	μ_{22}	ϕ_2
Total by genotype	θ_1	θ_2	1

In order to do so, we must first connect the concept of statistical signal size λ with some key quantities in association tests. While the term "signal size" likely sounds foreign to most practitioners, it is intimately linked with the concept of "effect sizes"—or odds ratios—in association studies, which are frequently estimated and reported in GWAS catalogs. Effect sizes, on the other hand, may be alien to some statisticians. In this section, we aim to bridge the two languages by characterizing the relationship between "signal size" and "odds-ratio" parameterizations in the special, but fairly common case of association tests on 2-by-2 contingency tables.

Recall the general setup of genetic association testing in Sect. 1.2, where one wants to detect the association between genetic variations at a specific location and the occurrence of a disease. An individual randomly drawn from the target population will have two (random) characteristics: a phenotype indicating whether the individual has the condition or is healthy (i.e., belonging to the Case group or the Control group), and a genotype that encodes the genetic variation in question. Table 1.1 in the introduction summarizes the *counts* for all phenotype–genotype combinations for the individuals in a given study sampled from the population. These counts may be assumed to follow a multinomial distribution, with probabilities given in Table 7.1.

Consider a 2-by-2 multinomial distribution with marginal probabilities of phenotypes (ϕ_1, ϕ_2) and genotypes (θ_1, θ_2). The *probability* table (as opposed to the table of multinomial *counts* in the introduction) is as follows.

The odds ratio (i.e., "effect size") is defined as the ratio of the phenotype frequencies between the two genotype variants,

$$R := \frac{\mu_{11}}{\mu_{21}} \bigg/ \frac{\mu_{12}}{\mu_{22}} = \frac{\mu_{11}\mu_{22}}{\mu_{12}\mu_{21}}. \tag{7.8}$$

The multinomial distribution is fully parametrized by the trio (θ_1, ϕ_1, R). Odds ratios further away from 1 indicate greater contrasts between the probabilities of outcomes. Independence between the genotypes and phenotypes would imply an odds ratio of one, and hence $\mu_{jk} = \phi_j\theta_k$, for all $j, k \in \{1, 2\}$.

For a sequence of local alternatives $\mu^{(1)}, \mu^{(2)}, \ldots$, such that $\sqrt{n}(\mu_{jk}^{(n)} - \phi_j\theta_k)$ converges to a constant table $\delta = (\delta_{jk})$, the chi-square test statistics converge in distribution to the non-central chi-squared distribution with non-centrality parameter

$$\lambda = \sum_{j=1}^{2} \sum_{k=1}^{2} \delta_{jk}^2/(\phi_j \theta_k).$$

See, e.g., Ferguson (2017). Hence, for large samples from a fixed distribution (μ_{ij}), the statistic is well approximated by a $\chi_1^2(\lambda)$ distribution, where

$$\lambda = n \sum_{j=1}^{2} \sum_{k=1}^{2} \frac{(\mu_{jk} - \phi_j \theta_k)^2}{\phi_j \theta_k}. \tag{7.9}$$

Power calculations therefore only depend on the μ_{jk}'s through $\lambda = nw^2$, where we define

$$w^2 := \lambda/n \tag{7.10}$$

to be the *signal size per sample*. Statistical power would be increasing in w^2 for fixed sample sizes.

The next proposition states that the statistical signal size per sample can be parametrized by the odds ratio and the marginals in the probability table.

Proposition 7.1 *Consider a 2-by-2 multinomial distribution with marginal distributions $(\phi_1, \phi_2 = 1 - \phi_2)$ and $(\theta_1, \theta_2 = 1 - \theta_1)$. Let signal size w^2 be defined as in (7.10), and odds ratio R be defined as in (7.8). If $R = 1$, we have $w^2 = 0$; if $R \in (0, 1) \cup (1, +\infty)$, then we have*

$$w^2(R) = \frac{1}{4A(R-1)^2} \left(B + CR - \sqrt{(B + CR)^2 - 4A(R-1)^2} \right)^2, \tag{7.11}$$

where $A = \phi_1 \theta_1 \phi_2 \theta_2$, $B = \phi_1 \theta_1 + \phi_2 \theta_2$, and $C = \phi_1 \theta_2 + \phi_2 \theta_1$.

Proof We parametrize the 2-by-2 multinomial distribution with the parameter δ,

$$\mu_{11} = \phi_1 \theta_1 + \delta, \quad \mu_{12} = \phi_1 \theta_2 - \delta, \quad \mu_{21} = \phi_2 \theta_1 - \delta, \quad \mu_{22} = \phi_2 \theta_2 + \delta. \tag{7.12}$$

By relabeling of categories, we may assume $0 < \theta_1, \phi_1 \le 1/2$ without loss of generality. Note that δ must lie within the range $[\delta_{\min}, \delta_{\max}]$, where

$$\delta_{\min} := \max\{-\phi_1 \theta_1, -\phi_2 \theta_2, \phi_1 \theta_2 - 1, \phi_2 \theta_1 - 1\} = -\phi_1 \theta_1,$$

and

$$\delta_{\max} := \min\{1 - \phi_1 \theta_1, 1 - \phi_2 \theta_2, \phi_1 \theta_2, \phi_2 \theta_1\} = \min\{\phi_1 \theta_2, \phi_2 \theta_1\},$$

in order for $\mu_{ij} \ge 0$ for all $i, j \in \{1, 2\}$. Under this parametrization, Relation (7.8) then becomes

$$R = \frac{\mu_{11} \mu_{22}}{\mu_{12} \mu_{21}} = \frac{\phi_1 \theta_1 \phi_2 \theta_2 + \delta(\phi_1 \theta_1 + \phi_2 \theta_2) + \delta^2}{\phi_1 \theta_1 \phi_2 \theta_2 - \delta(\phi_1 \theta_2 + \phi_2 \theta_1) + \delta^2} = \frac{A + \delta B + \delta^2}{A - \delta C + \delta^2}, \tag{7.13}$$

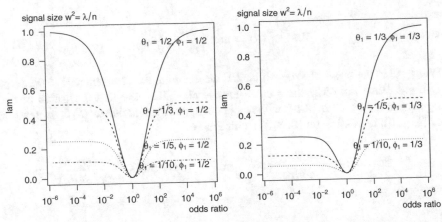

Fig. 7.1 Signal sizes per sample w^2 as functions of odds ratios in 2-by-2 multinomial distributions for selected genotype marginals in balanced (left) and unbalanced (right) designs; see Relation (7.11) in Proposition 7.1. For given marginal distributions, extreme odds ratios imply stronger statistical signals at a given sample size. However, the signal sizes are bounded above by constants that depend on the marginal distributions; see Relations (7.15) and (7.16)

which is one-to-one and increasing in δ on $(\delta_{\min}, \delta_{\max})$. Equation (7.10) becomes

$$w^2 = \sum_{i=1}^{2} \sum_{j=1}^{2} \frac{(\mu_{ij} - \phi_i \theta_j)^2}{\phi_i \theta_j} = \delta^2 \sum_i \sum_j \frac{1}{\phi_i \theta_j} = \frac{\delta^2}{\phi_1 \theta_1 \phi_2 \theta_2}, \qquad (7.14)$$

Solving for δ in (7.13), and plugging into the expression for signal size (7.14) yields Relation (7.11).

The other three cases ($1/2 \leq \theta_1, \phi_1 \leq 1$; $0 < \theta_1 \leq 1/2 \leq \phi_1 \leq 1$; and $0 \leq \phi_1 \leq 1/2 \leq \theta_1 \leq 1$) may be obtained similarly, or by appealing to the symmetry of the problem. $\qquad \square$

To understand Proposition 7.1, we illustrate Relation (7.11) for selected values of marginals θ_1 and ϕ_1 in Fig. 7.1. Observe in the figure that an odds ratio further away from one corresponds to stronger statistical signal per sample, ceteris paribus. However, this "valley" pattern is in general not symmetric around 1, except for balanced marginal distributions ($\phi_1 = 1/2$ or $\theta_1 = 1/2$). While the odds ratio R can be arbitrarily close to 0 or diverge to $+\infty$ for any marginal distribution, the signal sizes w^2 are bounded from above by constants that depend only on the marginals.

Corollary 7.1 *The signal size as a function of the odds ratio* $w^2(R)$ *is decreasing on* $(0, 1)$ *and increasing on* $(1, \infty)$*, with limits*

$$\lim_{R \to 0_+} w^2(R) = \min \left\{ \frac{\phi_1 \theta_1}{\phi_2 \theta_2}, \frac{\phi_2 \theta_2}{\phi_1 \theta_1} \right\}, \qquad (7.15)$$

and

$$\lim_{R \to +\infty} w^2(R) = \min\left\{\frac{\phi_1\theta_2}{\phi_2\theta_1}, \frac{\phi_2\theta_1}{\phi_1\theta_2}\right\}. \tag{7.16}$$

Proof As in the proof of Proposition 7.1, we examine the case where $0 < \theta_1, \phi_1 \le 1/2$, and leave the other three cases an exercise. Take the first derivative of the expression for w^2 in Eq. (7.14) with respect to δ, it is evident that $w^2(\delta)$ is decreasing on $[\delta_{\min}, 0)$, increasing on $(0, \delta_{\max}]$, with limits

$$\lim_{d \to \delta_{\min}} w^2(\delta) = \frac{\phi_1\theta_1}{\phi_2\theta_2}, \quad \text{and} \quad \lim_{d \to \delta_{\max}} w^2(\delta) = \min\left\{\frac{\phi_1\theta_2}{\phi_2\theta_1}, \frac{\phi_2\theta_1}{\phi_1\theta_2}\right\}.$$

$\square$

Corollary 7.1 immediately implies that balanced designs with roughly equal number of cases and controls are not necessarily the most informative.

For example, in a study where a third of the recruited subjects carry the genetic variant positively correlated with the trait (i.e., $\theta_1 = 1/3$), an unbalanced design with $\phi_1 = 1/3$ would maximize w^2 at large odds ratios. This unbalanced design is much more efficient compared to, say, a balanced design with $\phi_1 = 1/2$. In the first case, we have $w^2 \to 1$ as $R \to \infty$; whereas in the second design, $w^2 < 1/2$ no matter how large R is. This difference can also be seen by comparing the dashed curve ($\theta_1 = 1/3$, $\phi_1 = 1/2$) in the left panel of Fig. 7.1, with the solid curve ($\theta_1 = 1/3$, $\phi_1 = 1/3$) in the right panel of Fig. 7.1.

7.3 Optimal Study Designs and Rare Variants

For a study with a fixed budget, i.e., a fixed total number of subjects n, the researcher is free to choose the fraction of cases ϕ_1 to be included in the study. A natural question is how this budget should be allocated to maximize the statistical power of discovery, or equivalently, the signal sizes $\lambda = nw^2$.

In principal, Relation (7.11) can be optimized with respect to the fraction of cases ϕ_1 in order to find optimal designs, if θ_1 is known and held constant. In practice, this is not the case. While the fraction of cases can be controlled, the distributions of genotypes *in the study* are often unknown prior to data collection, and can change with the case-to-control ratio.

Fortunately, the conditional distributions of genotypes in the healthy control groups are often estimated by existing studies, and are made available by consortia such as the NHGRI-EBI GWAS catalog (MacArthur et al. 2016). We denote the conditional frequency of the first genetic variant in the control group as $(f, 1 - f)$, where

$$f := \mu_{21}/\phi_2 = \mu_{21}/(1 - \phi_1). \tag{7.17}$$

The multinomial probability is fully parametrized by the new trio: (f, ϕ_1, R).

Probabilities	Genotype Variant 1	Variant 2	Total by phenotype
Cases	$\frac{\phi_1 f R}{f R + 1 - f}$	$\frac{\phi_1 (1-f)}{f R + 1 - f}$	ϕ_1
Controls	$f(1 - \phi_1)$	$(1 - f)(1 - \phi_1)$	$1 - \phi_1$

Proposition 7.1 may also be re-stated in terms of the new parametrization.

Corollary 7.2 *In the 2-by-2 multinomial distribution with marginals* $(\phi_1, \phi_2 = 1 - \phi_1)$, *and conditional distribution of the variants in the control group* $(f, 1 - f)$, *Relation (7.11) holds with* $\theta_1 = \phi_1 f R / (f R + 1 - f) + f(1 - \phi_1)$ *and* $\theta_2 = 1 - \theta_1$.

The choice of ϕ_1 now has a practical solution.

Corollary 7.3 *In the context of Corollary 7.2, the optimal design* (ϕ_1^*, ϕ_2^*) *that maximizes the signal size per sample* w^2 *is prescribed by*

$$\phi_1^* = \frac{f R + 1 - f}{f R + 1 - f + \sqrt{R}}, \quad and \quad \phi_2^* = 1 - \phi_1^*. \tag{7.18}$$

Proof Using the parametrization in (7.12), we solve for δ in (7.13) to obtain

$$\delta = \frac{\phi_1 f R}{f R + 1 - f} - \left(\frac{\phi_1 f R}{f R + 1 - f} + f(1 - \phi_1) \right) \phi_1$$
$$= \frac{f(1 - f)\phi_1(1 - \phi_1)(R - 1)}{f R + 1 - f}. \tag{7.19}$$

Substituting (7.19) into the expression (7.14), after some simplification, yields

$$w^2 = \frac{f(1 - f)\phi_1(1 - \phi_1)(R - 1)^2}{[\phi_1 R + (1 - \phi_1)D][\phi_1 + (1 - \phi_1)D]}, \tag{7.20}$$

where $D = f R + 1 - f > 0$. Therefore, the derivative of (7.20) with respect to ϕ_1 is

$$\frac{dw^2}{d\phi_1} = \frac{f(1 - f)(R - 1)^2}{[\phi_1 R + (1 - \phi_1)D]^2 [\phi_1 + (1 - \phi_1)D]^2} \left[(D^2 - R)\phi_1^2 - 2D^2\phi_1 + D^2 \right]. \tag{7.21}$$

Further, we obtain the second derivative with respect to ϕ_1,

$$\frac{d^2 w^2}{d\phi_1^2} = h(R, f) \left[(\phi_1 - 1)D^2 - \phi_1 R \right], \tag{7.22}$$

where h is some function of (R, f) taking on strictly positive values.

Since $\left[(\phi_1 - 1)D^2 - \phi_1 R\right] < 0$, the second derivative (7.22) must be strictly negative on $[0, 1]$. This implies that the first derivative (7.21) is strictly decreasing on $[0, 1]$. Since the first derivative (7.21) is strictly positive at $\phi_1 = 0$, and strictly negative at $\phi_1 = 1$, it must have a unique zero between 0 and 1, and hence, the solution to $(D^2 - R)\phi_1^2 - 2D^2\phi_1 + D^2 = 0$ in the interval of $[0, 1]$ must be the maximizer of (7.20)—when $D^2 - R > 0$, the smaller of the two roots maximizes (7.20), and when $D^2 - R < 0$, it is the larger of the two. They share the same expression $D/(D + \sqrt{R})$, which coincides with (7.18). Finally, when $D^2 = R$, the only root $\phi_1^* = 1/2$, which also coincides with (7.18), is the maximizer of (7.20). $\square$

Of particular interest in the genetics literature are genetic variants with very low allele frequencies in the control group (i.e., $f \approx 0$), known as rare variants. In such cases, Eq. (7.18) can be approximated using the Taylor expansion,

$$\phi_1^* = \frac{1}{1 + \sqrt{R}} + \frac{(R - \sqrt{R})f}{1 + \sqrt{R}} + O(f^2). \tag{7.23}$$

To illustrate, for rare and adversarial factors ($f \approx 0$ and $R > 1$), the optimal ϕ_1^* is less than $1/2$. Therefore, for studies under a fixed budget, controls should constitute the majority of the subjects, in order to maximize power. On the other hand, for rare and protective factors ($f \approx 0$ and $R < 1$), the optimal ϕ_1^* is greater than $1/2$, and cases should be the majority.

7.4 Phase Transitions in Large-Scale Association Screening Studies

Returning to the problem of *high-dimensional* marginal screenings for categorical covariates, we explore the manifestation of the phase transition in the exact–approximate support recovery problem in the genetic context.

Recall Theorem 7.2 predicts that FWER and FNR can be simultaneously controlled in large dimensions if and only if

$$r = \frac{\lambda}{2 \log p} = \frac{w^2 n}{2 \log p} > 1. \tag{7.24}$$

Therefore, if we were to apply FWER-controlling procedures at low nominal levels (say, 5%), then the FNR would experience a phase transition in the following sense. If the signal size is strong enough, i.e.,

$$r > 1 \iff w^2 > \frac{2 \log p}{n}, \tag{7.25}$$

then the FNR can be close to 0; otherwise, FNR must be close to 1.

Using the parametric relationship described in Corollary 7.2 (and Proposition 7.1), the inequalities in (7.25) implicitly define regions of (f, R) where associations are discoverable with high power, for a given ϕ_1. Further, the boundary of such discoverable regions sharpens as dimensionality diverges. We illustrate this phase transition through a numerical example next.

Example 7.1 Consider association tests on 2×2 contingency tables at p locations as introduced in Sect. 1.2, where the counts follow a multinomial distribution parametrized by (f, R, ϕ_1) as in Sect. 7.3. Assume that the phenotype marginals are fixed at $\phi_1 = \phi_2 = 1/2$. Applying Bonferroni's procedure with nominal FWER at $\alpha = 5\%$ level, we can approximate the marginal power of association tests by

$$\mathbb{P}[\chi_1^2(\lambda) > \chi_{1,\alpha/p}^2], \tag{7.26}$$

where $\chi_{1,\alpha/p}^2$ is the upper (α/p)-quantile of a central chi-squared distribution with 1 degree of freedom. We calculate this marginal power as a function of the parameters (f, R) in three scenarios:

- $p = 4, n = 3 \times 10^4$
- $p = 10^2, n = 1 \times 10^5$
- $p = 10^6, n = 3 \times 10^6$

and visualize the results as heatmaps[2] (referred to as OR-RAF diagrams) in Fig. 7.2. These parameter values are chosen so that $\log(p)/n$ are roughly constant (around 4.6×10^{-5}).

We also overlay "equi-signal" curves, i.e., functions implicitly defined by the equations $r = c$ for a range of c (dashed curves), and highlight the predicted boundary of phase transition for the exact–approximate support recovery problem $r = 1$ (red curves). The change in marginal power clearly sharpens around the predicted boundary $r = 1$ as dimensionality diverges.

Remark 7.2 In an attempt to find empirical evidence of our theoretical predictions, we chart the genetic variants associated with breast cancer, discovered in a 2017 study by Michailidou et al. (2017) in an OR-RAF diagram. The estimated risk allele frequencies (f) and odds ratios (R) are taken from the NHGRI-EBI GWAS catalog MacArthur et al. (2016), and plotted against a power heatmap calculated according to the reported sample sizes. See lower-right panel of Fig. 7.2.

It is tempting to believe, on careless inspection, that roughly *all* discovered associations fall inside the high power region of the diagram, therefore demonstrating the phase transition in statistical power. Unfortunately, the estimates here are subject to survival bias—the study in fact uses the same dataset for *both* support estimation and parameter estimation, without adjusting the latter for the selection process. The seemingly striking agreement between the power calculations and the estimated effects of

---

[2] Since genetic variants can always be relabelled such that Variant 1 is positively associated with cases, we only produce part of the diagram where $R > 1$. Sample sizes marked in the figure are adjusted by a factor of $1/2$, to reflect the genetic context where a pair of alleles are measured for every individual at every genomic location.

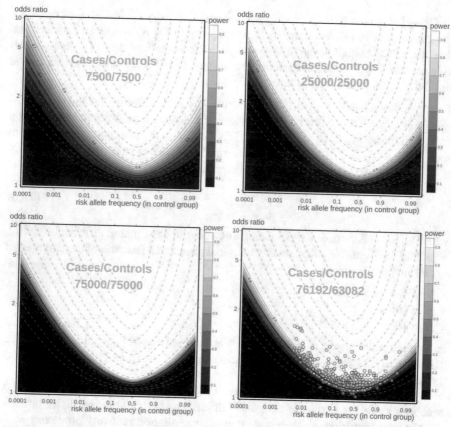

Fig. 7.2 The OR-RAF diagram visualizing the marginal power of discovery in genetic association studies, after applying Bonferroni's procedure with nominal FWER at 5% level. Sample sizes are marked in each panel, and the problem dimensions are, respectively, $p = 4$ (upper-left), $p = 10^2$ (upper-right), and $p = 10^6$ (lower-left), so that $n/\log p$ are roughly constant. Red curves mark the boundaries ($r = 1$) of the phase transition for the exact-approximate support recovery problem; dashed curves are the equi-signal (equi-power) curves. The phase transition in signal sizes λ translates into the phase transition in terms of (f, R), and sharpens as $p \to \infty$; see Example 7.1. In the lower right panel, we visualize discovered associations (blue circles) in a recent GWA study (Michailidou et al. 2017); the estimated odds ratios and risk allele frequencies are subject to survival bias and should not be taken at their face values; see Remark 7.2

reported associations *should not* be taken as conclusive evidence for the validity of our theory. We conjecture, as the theory predicts, that accurate and unbiased parameter estimates from an independent replication will still place the associations in the high power region of the diagram.

Finally, we demonstrate with an example how results in Sects. 7.1 and 7.2 may be used for planning prospective association studies.

Example 7.2 In a GWAS with $p = 10^6$ genomic marker locations, researchers wish to locate genetic associations with the trait of interest. Specifically, they wish to maximize power in the region where genetic variants have risk allele frequencies of 0.01 and odds ratios of 1.2. By Corollary 7.3, the optimal design has a fraction of cases $\phi^* = 0.478$, yielding the statistical signal size per sample $w^2 \approx 9.00 \times 10^{-5}$ according to Corollary 7.2.

If we wish to achieve exact–approximate support recovery in the sense of (2.25), Theorem 7.2 predicts that the signal size parameter r has to be at least $f_{EA}(\beta) = 1$. This signal size calls for a sample size of $n = \lambda/w^2 = 2r \log(p)/w^2 \approx 307, 011$. In a typical GWAS, a pair of alleles are sequenced for every marker location, bringing the required number of subjects in the study to $n/2 \approx 153, 509$.

In comparison, a more accurate power calculation directly using (7.26) predicts that $n/2 = 165, 035$ subjects are needed, under the set of parameters ($p = 10^6$, $f = 0.01, R = 1.2$) and FWER $= 0.05$, FNR $= 0.5$; this is 7% higher than our crude asymptotic approximation. In general, we recommend using the more precise calculations over the back-of-the-envelope asymptotics for planning prospective studies and performing systematic reviews; a user-friendly web application implementing the more precise approximations is provided in Gao et al. (2019). Nevertheless, the theoretical results on phase transitions generate simple, accurate, and powerful insights that cannot be easily derived from numerical calculations.

7.5 Numerical Illustrations of the Phase Transitions in Chi-Square Models

We illustrate with simulations the phase transition phenomena in the chi-square model, and compare numerically the required signal sizes in support recovery problems between the two types of alternatives in the additive error model.

7.5.1 Exact Support Recovery

The sparsity of the signal vectors in the experiments are parametrized as in (7.2). Signal sizes are assumed equal with magnitude $\lambda(i) = 2r \log p$ for $i \in S$. We estimate the support set S using Bonferroni's procedure with nominal FWER level set at $1/(5\log p)$. The nominal FWER levels vanishes slowly, in line with the assumptions in Theorem 7.1. Experiments were repeated 1000 times at each of the 400 sparsity-signal-size combinations, for dimension $p = 10^4$.

The empirical probabilities of exact support recovery under Bonferroni's procedure are shown in Fig. 7.3. The numerical results suggest good accuracy of the predicted boundaries in high-dimensions (left panels of Fig. 7.3).

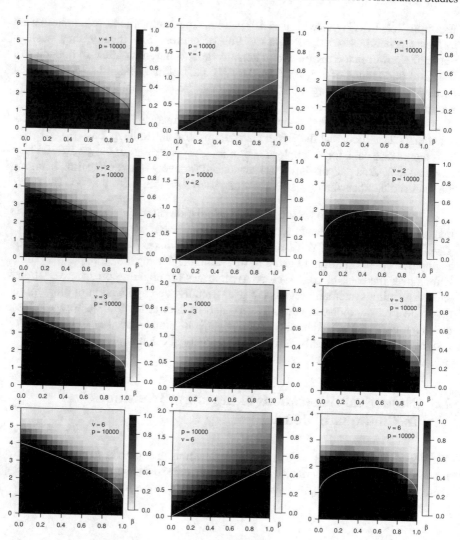

Fig. 7.3 The empirical risks of exact, approximate, and approximate–exact support recovery (left to right) in the chi-squared model (1.3) with Bonferroni's procedure and the Benjamini–Hochberg procedure. We display results as a heatmap for $\nu = 1, 2, 3, 6$ (first to last row) at dimension $p = 10^4$, for a grid of sparsity levels β and signal sizes r. The experiments were repeated 1000 times for each sparsity-signal size combination; darker color indicates higher probability of exact support recovery. Numerical results are in general agreement with the boundaries described in Theorem 7.1; for large ν's, the phase transitions take place somewhat above the predicted boundaries

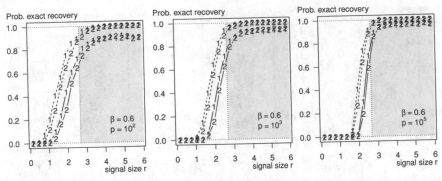

Fig. 7.4 The empirical probability of exact support recovery of Bonferroni's procedure (solid curves) and the oracle procedure (dashed curves) in the chi-squared model with one degree of freedom (marked '2') in the additive Gaussian error model and under one-sided alternatives (marked '1'). We simulate at dimensions $p = 10^2, 10^3, 10^5$ (left to right) for a grid of signal sizes r and sparsity level $\beta = 0.6$. The experiments were repeated 1000 times for each method-model-signal-size combination. Numerical results show evidence of convergence to the 0–1 law as predicted by Theorem 7.1; regions where asymptotically exact support recovery can be achieved are shaded in grey. The difference in power between Bonferroni's procedure and the oracle procedure, as well as in the two types of alternatives both decrease as dimensionality increases

We conduct further experiments to examine the optimality claims in Theorem 7.1 by comparing with the oracle procedure with thresholds $t_p = \min_{i \in S} x(i)$. We also examine the claims in Sect. 7.1.5, and compare the one-sided alternatives in Gaussian additive models with the two-sided alternatives (or equivalently, the chi-square model with $\nu = 1$). We apply Bonferroni's procedure and the oracle thresholding procedure in both settings.

The experiments were repeated 1000 times for a grid of signal size values ranging from $r = 0$ to 6, and for dimensions $10^2, 10^3$, and 10^5. Results of the experiments, shown in Fig. 7.4, suggest vanishing difference between difficulties of two-sided vs one-sided alternatives in the additive error models, as well as vanishing difference between the powers of Bonferroni's procedures and the oracle procedures as $p \to \infty$.

7.5.2 Approximate, and Approximate–Exact Support Recovery

Similar experiments are conducted to examine the optimality claims in Theorem 7.3, and in Sect. 7.1.5. We define an oracle thresholding procedure for approximate support recovery, where the threshold is chosen to minimize the empirical risk. That is,

$$t_p(x, S) \in \arg\min_{t \in \mathbb{R}} \frac{|\widehat{S}(t) \setminus S|}{\max\{|\widehat{S}(t)|, 1\}} + \frac{|S \setminus \widehat{S}(t)|}{\max\{|S|, 1\}},$$

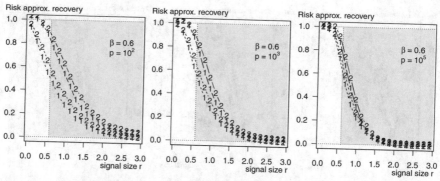

Fig. 7.5 The empirical risk of approximate support recovery of Benjamini–Hochberg's procedure (solid curves) and the oracle procedure (dashed curves) in the chi-squared model with one degree of freedom (marked '2') and in the additive Gaussian error model under one-sided alternatives (marked '1'). We simulate at dimensions $p = 10^2, 10^3, 10^5$ (left to right) for a grid of signal sizes r and sparsity level $\beta = 0.6$. The experiments were repeated 1000 times for each method-model-signal-size combination. Numerical results show evidence of convergence to the 0–1 law as predicted by Theorem 7.3; regions where asymptotically approximate support recovery can be achieved are shaded in grey. The difference in risks between Benjamini–Hochberg's procedure and the oracle procedure, as well as in the two types of alternatives, both decrease as dimensionality increases

where $\widehat{S}(t) = \{i \mid x(i) \geq t\}$; in implementation, we only need to scan the values of observations $t \in \{x(1), \ldots, x(p)\}$. The nominal FDR level for the BH procedure is set at $1/(5\log p)$, therefore slowly vanishing, in line with the assumptions in Theorem 7.3; all other parameters are identical to that in the experiments for exact support recovery in Sect. 7.5.1. The results of the experiments are shown in Fig. 7.5 and in the middle column of Fig. 7.3.

We also examine the boundary described in Theorem 7.2. Experimental settings are identical to that in the experiments for approximate support recovery. We compare the performance of the BH procedure with an oracle procedure with threshold

$$t_p(x, S) \in \min_{i \in S} x(i),$$

and visualize the results of the experiments in the right column of Fig. 7.3. Notice that the BH procedure sets its threshold somewhat higher than the oracle, especially for small β's. The empirical risk of the oracle procedure (not shown here in the interest of space) follows much more closely the predicted boundary (7.7).

Appendix A
Additional Proofs

We review some properties of the chi-square distributions in Sect. A.1, before presenting the proofs of the main theorems on phase transitions in Sects. A.2, A.3, and A.4.

A.1 Auxiliary Facts of Chi-Square Distributions

We shall recall, and establish, some auxiliary facts about chi-square distributions. These facts will be used in the proofs of Theorems 7.1 and 7.3.

Lemma A.1 (Rapid variation of chi-square distribution tails) *The central chi-square distribution with v degrees of freedom has rapidly varying tails. That is,*

$$\lim_{x \to \infty} \frac{\mathbb{P}[\chi_v^2(0) > tx]}{\mathbb{P}[\chi_v^2(0) > x]} = \begin{cases} 0, & t > 1 \\ 1, & t = 1 \\ \infty, & 0 < t < 1 \end{cases}, \tag{A.1}$$

where we overloaded the notation $\chi_v^2(0)$ to represent a random variable with the chi-square distribution.

Proof (*Lemma A.1*) When $v = 1$, the chi-square distribution reduces to a squared Normal, and (A.1) follows from the rapid variation of the standard Normal distribution. For $v \geq 2$, we recall the following bound on tail probabilities (see, e.g., Inglot 2010),

$$\frac{1}{2}\mathcal{E}_v(x) \leq \mathbb{P}[\chi_v^2(0) > x] \leq \frac{x}{(x - v + 2)\sqrt{\pi}}\mathcal{E}_v(x), \quad v \geq 2, \ x > v - 2,$$

© The Author(s), under exclusive license to Springer Nature Switzerland AG 2021
Z. Gao and S. Stoev, *Concentration of Maxima and Fundamental
Limits in High-Dimensional Testing and Inference*,
SpringerBriefs in Probability and Mathematical Statistics,
https://doi.org/10.1007/978-3-030-80964-5

where $\mathcal{E}_v(x) = \exp\left\{-\frac{1}{2}[(x - v - (v - 2)\log(x/v) + \log v]\right\}$. Therefore, we have

$$\frac{(x - v + 2)\sqrt{\pi}}{2x} \frac{\mathcal{E}_v(tx)}{\mathcal{E}_v(x)} \leq \frac{\mathbb{P}[\chi_v^2(0) > tx]}{\mathbb{P}[\chi_v^2(0) > x]} \leq \frac{2tx}{(tx - v + 2)\sqrt{\pi}} \frac{\mathcal{E}_v(tx)}{\mathcal{E}_v(x)},$$

where $\mathcal{E}_v(tx)/\mathcal{E}_v(x) = \exp\{-\frac{1}{2}[(t - 1)x - (v - 2)\log t]\}$ converges to 0 or ∞ depending on whether $t > 1$ or $0 < t < 1$. The case where $t = 1$ is trivial. $\qquad\square$

Lemma A.1 and Proposition 2.2 yield the following Corollary.

Corollary A.1 *Maxima of independent observations from central chi-square distributions with v degrees of freedom are relatively stable. Specifically, let $\epsilon_p = \left(\epsilon_p(i)\right)_{i=1}^p$ be independently and identically distributed (iid) $\chi_v^2(0)$ random variables. Then the triangular array $\mathcal{E} = \{\epsilon_p, p \in \mathbb{N}\}$ has relatively stable (RS) maxima in the sense of (2.38).*

Lemma A.2 (Stochastic monotonicity) *The non-central chi-square distribution is stochastically monotone in its non-centrality parameter. Specifically, for two non-central chi-square distributions both with v degrees of freedom, and non-centrality parameters $\lambda_1 \leq \lambda_2$, we have $\chi_v^2(\lambda_1) \overset{d}{\leq} \chi_v^2(\lambda_2)$. That is,*

$$\mathbb{P}[\chi_v^2(\lambda_1) \leq t] \geq \mathbb{P}[\chi_v^2(\lambda_2) \leq t], \quad \text{for any } t \geq 0. \tag{A.2}$$

where we overloaded the notation $\chi_v^2(\lambda)$ to represent a random variable with the chi-square distribution with non-centrality parameter λ and degree-of-freedom parameter v.

Proof *(Lemma A.2)* Recall that non-central chi-square distributions can be written as sums of $v - 1$ standard normal random variables and a non-central normal random variable with mean $\sqrt{\lambda}$ and variance 1,

$$\chi_v^2(\lambda) \overset{d}{=} Z_1^2 + \ldots + Z_{v-1}^2 + (Z_v + \sqrt{\lambda})^2.$$

Therefore, it suffices to show that $\mathbb{P}[(Z + \sqrt{\lambda})^2 \leq t]$ is non-increasing in λ for any $t \geq 0$, where Z is a standard normal random variable. We rewrite this expression in terms of standard normal probability function Φ,

$$\mathbb{P}[(Z + \sqrt{\lambda})^2 \leq t] = \mathbb{P}[-\sqrt{\lambda} - \sqrt{t} \leq Z \leq -\sqrt{\lambda} + \sqrt{t}]$$
$$= \Phi(-\sqrt{\lambda} + \sqrt{t}) - \Phi(-\sqrt{\lambda} - \sqrt{t}). \tag{A.3}$$

The derivative of the last expression (with respect to λ) is

$$\frac{1}{2\sqrt{\lambda}}\left(\phi(\sqrt{\lambda} + \sqrt{t}) - \phi(\sqrt{\lambda} - \sqrt{t})\right) = \frac{1}{2\sqrt{\lambda}}\left(\phi(\sqrt{\lambda} + \sqrt{t}) - \phi(\sqrt{t} - \sqrt{\lambda})\right), \tag{A.4}$$

where ϕ is the density of the standard normal distribution. Notice that we have used the symmetry of ϕ around 0 in the last expression.

Since $0 \leq \max\{\sqrt{\lambda} - \sqrt{t}, \sqrt{t} - \sqrt{\lambda}\} < \sqrt{t} + \sqrt{\lambda}$ when $t > 0$, by monotonicity of the normal density on $(0, \infty)$, we conclude that the derivative (A.4) is indeed negative. Therefore, (A.3) is decreasing in λ, and (A.2) follows for $t > 0$. For $t = 0$, equality holds in (A.2) with both probabilities being 0. $\qquad\square$

Finally, we derive asymptotic expressions for chi-square quantiles.

Lemma A.3 (Chi-square quantiles) *Let F be the central chi-square distributions with ν degrees of freedom, and let $u(y)$ be the $(1-y)$-th generalized quantile of F, i.e.,*

$$u(y) = F^{\leftarrow}(1-y). \tag{A.5}$$

Then

$$u(y) \sim 2\log(1/y), \quad as \ y \to 0. \tag{A.6}$$

Proof (*Lemma* A.3) The case where $\nu = 1$ follows from the well-known formula for Normal quantiles (see, e.g., Proposition 1.1 in Gao and Stoev 2020)

$$F^{\leftarrow}(1-y) = \Phi^{\leftarrow}(1 - y/2) \sim \sqrt{2\log(2/y)} \sim \sqrt{2\log(1/y)}.$$

The case where $\nu \geq 2$ follows from the following estimates of high quantiles of chi-square distributions (see, e.g., Inglot 2010),

$$\nu + 2\log(1/y) - 5/2 \leq u(y) \leq \nu + 2\log(1/y) + 2\sqrt{\nu\log(1/y)}, \quad \text{for all } y \leq 0.17,$$

where both the lower and upper bound are asymptotic to $2\log(1/y)$. $\qquad\square$

A.2 Proof of Theorem 7.1

Proof (*Theorem* 7.1) We first prove the sufficient condition. The Bonferroni procedure sets the threshold at $t_p = F^{\leftarrow}(1 - \alpha/p)$, which, by Lemma A.3, is asymptotic to $2\log p - 2\log \alpha$. By the assumption on α in (3.17), for any $\delta > 0$, we have $p^{-\delta} = o(\alpha)$. Therefore, we have $-\log \alpha \leq \delta \log p$ for large p, and

$$1 \leq \limsup_{p \to \infty} \frac{2\log p - 2\log \alpha}{2\log p} \leq 1 + \delta,$$

for any $\delta > 0$. Hence, $t_p \sim 2\log p$.

The condition $\underline{r} > f_E(\beta)$ implies, after some algebraic manipulation, $\sqrt{\underline{r}} - \sqrt{1 - \beta} > 1$. Therefore, we can pick $q > 1$ such that

$$\sqrt{r} - \sqrt{1-\beta} > \sqrt{q} > 1. \tag{A.7}$$

Setting the $t^* = t_p^* = 2q \log p$, we have $t_p < t_p^*$ for large p.

On the one hand, FWER $= 1 - \mathbb{P}[\widehat{S}_p \subseteq S_p]$ vanishes under the Bonferroni procedure with $\alpha \to 0$. On the other hand, for large p, the probability of no missed detection is bounded from below by

$$\mathbb{P}[\widehat{S}_p \supseteq S_p] = \mathbb{P}[\min_{i \in S} x(i) \geq t_p] \geq \mathbb{P}[\min_{i \in S} x(i) \geq t^*] \geq 1 - p^{1-\beta} \mathbb{P}[\chi_\nu^2(\Delta) < t^*],$$
$$\tag{A.8}$$

where we have used the fact that signal sizes are bounded below by Δ, and the stochastic monotonicity of chi-square distributions (Lemma A.2) in the last inequality. Writing

$$\chi_\nu^2(\Delta) \stackrel{\mathrm{d}}{=} Z_1^2 + \ldots + Z_{\nu-1}^2 + (Z_\nu + \sqrt{\Delta})^2$$

where Z_i's are iid standard normal variables, we have

$$\mathbb{P}[\chi_\nu^2(\Delta) < t^*] \leq \mathbb{P}[(Z_\nu + \sqrt{\Delta})^2 < t^*] = \mathbb{P}[|Z_\nu + \sqrt{\Delta}| < \sqrt{t^*}]$$
$$\leq \mathbb{P}\left[Z_\nu < -\sqrt{\Delta} + \sqrt{t^*}\right]$$
$$= \mathbb{P}\left[Z_\nu < \sqrt{2 \log p}\left(\sqrt{q} - \sqrt{r}\right)\right]. \tag{A.9}$$

By our choice of q in (A.7), the last probability in (A.9) can be bounded from above by

$$\mathbb{P}\left[Z_\nu < -\sqrt{2(1-\beta)\log p}\right] \sim \frac{\phi\left(-\sqrt{2(1-\beta)\log p}\right)}{\sqrt{2(1-\beta)\log p}}$$
$$= \frac{1}{\sqrt{2(1-\beta)\log p}} p^{-(1-\beta)},$$

where the first line uses Mill's ratio for Gaussian distributions (see Sect. 2.7 and Relation (2.45)). This, combined with (A.8), completes the proof of the sufficient condition for the Bonferroni's procedure.

Under the assumption of independence, Sidák's, Holm's, and Hochberg's procedures are strictly more powerful than Bonferroni's procedure, while controlling FWER at the nominal levels. Therefore, the risks of exact support recovery for these procedures also vanishes. This completes the proof for the first part of Theorem 7.1.

We now show the necessary condition. We first normalize the maxima by the chi-square quantiles $u_p = F^{\leftarrow}(1 - 1/p)$, where F is the distribution of a (central) chi-square random variable,

$$\mathbb{P}[\widehat{S}_p = S_p] \leq \mathbb{P}\left[M_{S^c} < t_p \leq m_S\right] \leq \mathbb{P}\left[\frac{M_{S^c}}{u_p} < \frac{m_S}{u_p}\right], \tag{A.10}$$

where $M_{S^c} = \max_{i \in S^c} x(i)$ and $m_S = \min_{i \in S} x(i)$. By the relative stability of chi-square random variables (Corollary A.1), we know that $M_{S^c}/u_{|S^c|} \to 1$ in probability. Further, using the expression for u_p (Lemma A.3), we obtain

$$\frac{u_{p-p^{1-\beta}}}{u_p} \sim \frac{2\log(p - p^{1-\beta})}{2\log p} = \frac{\log p + \log(1 - p^{-\beta})}{\log p} \sim 1.$$

Therefore, the left-hand-side of the last probability in (A.10) converges to 1,

$$\frac{M_{S^c}}{u_p} = \frac{M_{S^c}}{u_{p-p^{1-\beta}}} \frac{u_{p-p^{1-\beta}}}{u_p} \xrightarrow{\mathbb{P}} 1. \tag{A.11}$$

Meanwhile, for any $i \in S$, by Lemma A.2 and the fact that signal sizes are bounded above by $\overline{\Delta}$, we have,

$$\chi_\nu^2(\lambda(i)) \stackrel{\mathrm{d}}{\leq} \chi_\nu^2(\overline{\Delta}) \stackrel{\mathrm{d}}{=} Z_1^2 + \ldots + Z_{\nu-1}^2 + \left(Z_\nu + \sqrt{\overline{\Delta}}\right)^2.$$

Dividing through by u_p, and taking minimum over S, we obtain

$$\frac{m_S}{u_p} = \min_{i \in S} \frac{\chi_\nu^2(\lambda(i))}{u_p} \stackrel{\mathrm{d}}{\leq} \min_{i \in S}\left\{\frac{Z_1^2(i) + \ldots + Z_{\nu-1}^2(i)}{u_p} + \frac{(Z_\nu(i) + \sqrt{\overline{\Delta}})^2}{u_p}\right\}. \tag{A.12}$$

Let $i^\dagger = i_p^\dagger$ be the index minimizing the second term in (A.12), i.e.,

$$i^\dagger := \arg\min_{i \in S} \frac{(Z_\nu(i) + \sqrt{\overline{\Delta}})^2}{u_p} = \arg\min_{i \in S} f_p(Z_\nu(i)), \tag{A.13}$$

where $f_p(x) := (x + \sqrt{\overline{\Delta}})^2/(2\log p)$. We shall first show that

$$\mathbb{P}[f_p(Z_\nu(i^\dagger)) < 1 - \delta] \to 1, \tag{A.14}$$

for some small $\delta > 0$. On the one hand, we know (by solving a quadratic inequality) that

$$f_p(x) < 1 - \delta \iff \frac{x}{\sqrt{2\log p}} \in (-(\sqrt{\bar{r}} + \sqrt{1 - \delta}), -(\sqrt{\bar{r}} - \sqrt{1 - \delta})). \tag{A.15}$$

On the other hand, we know (by the relative stability of iid Gaussians, recall Sect. 2.7) that

$$\frac{\min_{i \in S} Z_\nu(i)}{\sqrt{2\log p}} \to -\sqrt{1 - \beta} \quad \text{in probability.} \tag{A.16}$$

Further, by the assumption on the signal sizes $\bar{r} < (1 + \sqrt{1 - \beta})^2$, we have,

$$-(\sqrt{\bar{r}} + 1) < -1 < -\sqrt{1 - \beta} < -(\sqrt{\bar{r}} - 1).$$

Therefore we can picking a small $\delta > 0$ such that

$$-(\sqrt{\bar{r}} + 1) < -(\sqrt{\bar{r}} + \sqrt{1 - \delta}) < -\sqrt{1 - \beta} < -(\sqrt{\bar{r}} - \sqrt{1 - \delta}) < -(\sqrt{\bar{r}} - 1).$$
(A.17)

Combining (A.15), (A.16), and (A.17), we obtain

$$\mathbb{P}\left[\min_{i \in S} f_p(Z_\nu(i)) < 1 - \delta\right] = \mathbb{P}\left[f_p(Z_\nu(i^\dagger)) < 1 - \delta\right]$$

$$\geq \mathbb{P}\left[f_p\left(\min_{i \in S} Z_\nu(i)\right) < 1 - \delta\right] \to 1,$$

and we arrive at (A.14). As a corollary, since $u_p \sim 2 \log p$, it follows that

$$\mathbb{P}\left[\min_{i \in S} \frac{(Z_\nu(i) + \sqrt{\Delta})^2}{u_p} < 1 - \delta\right] \to 1. \tag{A.18}$$

Finally, by independence between $Z_1^2(i) + \ldots + Z_{\nu-1}^2(i)$ and $(Z_\nu^2(i) + \sqrt{\overline{\Delta}})^2$, and the fact that $i^\dagger$ is a function of only the latter, we have

$$Z_1^2(i^\dagger) + \ldots + Z_{\nu-1}^2(i^\dagger) \overset{d}{=} Z_1^2(i) + \ldots + Z_{\nu-1}^2(i) \quad \text{for all } i \in S.$$

Therefore, $Z_1^2(i^\dagger) + \ldots + Z_{\nu-1}^2(i^\dagger) = O_\mathbb{P}(1)$, and

$$\frac{Z_1^2(i^\dagger) + \ldots + Z_{\nu-1}^2(i^\dagger)}{u_p} \to 0 \quad \text{in probability.} \tag{A.19}$$

Together, (A.18) and (A.19) imply that

$$\mathbb{P}\left[\frac{m_S}{u_p} < 1 - \delta\right] \geq \mathbb{P}\left[\min_{i \in S} \left\{\frac{Z_1^2(i) + \ldots + Z_{\nu-1}^2(i)}{u_p} + \frac{(Z_\nu(i) + \sqrt{\Delta})^2}{u_p}\right\} < 1 - \delta\right]$$

$$\geq \mathbb{P}\left[\frac{Z_1^2(i^\dagger) + \ldots + Z_{\nu-1}^2(i^\dagger)}{u_p} + \frac{(Z_\nu(i^\dagger) + \sqrt{\Delta})^2}{u_p} < 1 - \delta\right] \to 1. \tag{A.20}$$

In view of (A.10), (A.11), and (A.20), we conclude that exact recovery cannot succeed with any positive probability. The proof of the necessary condition is complete. $\square$

A.3 Proof of Theorem 7.3

We first show the necessary condition. That is, when $\bar{r} < \beta$, no thresholding procedure is able to achieve approximate support recovery.

The proof follows the ideas in Arias-Castro and Chen (2017), and is very similar to the proof of Theorem 3.3. One could in principle obtain the proofs in this section by referencing arguments that have appeared in Chap. 3. We choose to present the proof here in full for completeness.

Proof (*Necessary condition in Theorem 7.3*) Denote the distributions of $\chi_\nu^2(0)$, $\chi_\nu^2(\Delta)$ and $\chi_\nu^2(\overline{\Delta})$ as F_0, $F_{\underline{a}}$, and $F_{\overline{a}}$ respectively.

Recall that thresholding procedures are of the form

$$\widehat{S}_p = \{ i \mid x(i) > t_p(x) \}.$$

Denote $\widehat{S} := \{ i \mid x(i) > t_p(x) \}$, and $\widehat{S}(u) := \{ i \mid x(i) > u \}$. For any threshold $u \geq t_p$ we must have $\widehat{S}(u) \subseteq \widehat{S}$, and hence

$$\text{FDP} := \frac{|\widehat{S} \setminus S|}{|\widehat{S}|} \geq \frac{|\widehat{S} \setminus S|}{|\widehat{S} \cup S|} = \frac{|\widehat{S} \setminus S|}{|\widehat{S} \setminus S| + |S|} \geq \frac{|\widehat{S}(u) \setminus S|}{|\widehat{S}(u) \setminus S| + |S|}. \tag{A.21}$$

On the other hand, for any threshold $u \leq t_p$ we must have $\widehat{S}(u) \supseteq \widehat{S}$, and hence

$$\text{NDP} := \frac{|S \setminus \widehat{S}|}{|S|} \geq \frac{|S \setminus \widehat{S}(u)|}{|S|}. \tag{A.22}$$

Since either $u \geq t_p$ or $u \leq t_p$ must take place, putting (A.21) and (A.22) together, we have

$$\text{FDP} + \text{NDP} \geq \frac{|\widehat{S}(u) \setminus S|}{|\widehat{S}(u) \setminus S| + |S|} \wedge \frac{|S \setminus \widehat{S}(u)|}{|S|}, \tag{A.23}$$

for any u. Therefore it suffices to show that for a suitable choice of u, the RHS of (A.23) converges to 1 in probability; the desired conclusion on FDR and FNR follows by the dominated convergence theorem.

Let $t^* = 2q \log p$ for some fixed q, we obtain an estimate of the tail probability

$$\overline{F_0}(t^*) = \mathbb{P}[\chi_\nu^2(0) > t^*] = \frac{2^{-\nu/2}}{\Gamma(\nu/2)} \int_{2q \log p}^{\infty} x^{\nu/2-1} e^{-x/2} dx$$

$$\sim \frac{2^{-\nu/2}}{\Gamma(\nu/2)} 2 \, (2q \log p)^{\nu/2-1} \, p^{-q}. \tag{A.24}$$

where $a_p \sim b_p$ is taken to mean $a_p/b_p \to 1$; this tail estimate was also obtained in Donoho and Jin (2004). Observe that $|\widehat{S}(t^*) \setminus S|$ has distribution $\text{Binom}(p - s, \overline{F_0}(t^*))$ where $s = |S|$, denote $X = X_p := |\widehat{S}(t^*) \setminus S|/|S|$, and we have

$$\mu := \mathbb{E}[X] = \frac{(p - s)\overline{F_0}(t^*)}{s}, \quad \text{and} \quad \text{Var}(X) = \frac{(p - s)\overline{F_0}(t^*)F_0(t^*)}{s^2} \le \mu/s.$$

Therefore for any $M > 0$, we have, by Chebyshev's inequality,

$$\mathbb{P}[X < M] \le \mathbb{P}[|X - \mu| > \mu - M] \le \frac{\mu/s}{(\mu - M)^2} = \frac{1/(\mu s)}{(1 - M/\mu)^2}. \tag{A.25}$$

Now, from the expression of $\overline{F_0}(t^*)$ in (A.24), we obtain

$$\mu = (p^\beta - 1)\overline{F_0}(t^*) \sim \frac{2^{1-\nu/2}}{\Gamma(\nu/2)} (2q \log p)^{\nu/2-1} p^{\beta-q}.$$

Since $\bar{r} < \beta$, we can pick q such that $\bar{r} < q < \beta$. In turn, we have $\mu \to \infty$, as $p \to \infty$. Therefore the last expression in (A.25) converges to 0, and we conclude that $X \to \infty$ in probability, and hence

$$\frac{|\widehat{S}(t^*) \setminus S|}{|\widehat{S}(t^*) \setminus S| + |S|} = \frac{X}{X + 1} \to 1 \quad \text{in probability}. \tag{A.26}$$

On the other hand, we show that with the same choice of $u = t^*$,

$$\frac{|S \setminus \widehat{S}(t^*)|}{|S|} \to 1 \quad \text{in probability}. \tag{A.27}$$

By the stochastic monotonicity of chi-square distributions (Lemma A.2), the probability of missed detection for each signal is lower bounded by $\mathbb{P}[\chi^2_\nu(\lambda_i) \le t^*] \ge F_{\bar{a}}(t^*)$. Therefore, $|S \setminus \widehat{S}(t^*)| \overset{d}{\ge} \text{Binom}(s, F_{\bar{a}}(t^*))$, and it suffices to show that $F_{\bar{a}}(t^*)$ converges to 1. This is indeed the case, since

$$F_{\bar{a}}(t^*) = \mathbb{P}[Z_1^2 + \ldots + Z_\nu^2 + 2\sqrt{2\bar{r} \log p}\, Z_\nu + 2\bar{r} \log p \le 2q \log p]$$
$$\ge \mathbb{P}[Z_1^2 + \ldots + Z_\nu^2 \le (q - \bar{r}) \log p,\ 2\sqrt{2\bar{r} \log p}\, Z_\nu \le (q - \bar{r}) \log p],$$

and both events in the last line have probability going to 1 as $p \to \infty$. The necessary condition is shown. $\square$

We now turn to the sufficient condition. That is, when $\underline{r} > \beta$, the Benjamini–Hochberg procedure with slowly vanishing FDR levels achieves asymptotic approximate support recovery. The structure for the proof of sufficient condition follows that of Theorem 2 in Arias-Castro and Chen (2017).

Proof (*Sufficient condition in Theorem 7.3*) The FDR vanishes by our choice of α and the FDR-controlling property of the BH procedure. It only remains to show that FNR also vanishes.

To do so we compare the FNR under the alternative specified in Theorem 7.3 to one with all of the signal sizes equal to $\underline{\Delta}$. Let $x(i)$ be vectors of independent observations with $p - s$ nulls having $\chi_\nu^2(0)$ distributions, and s signals having $\chi_\nu^2(\underline{\Delta})$ distributions. By Lemma 3.2, it suffices to show that the FNR under the BH procedure in this setting vanishes.

Let $\widehat{G}$ denote the empirical survival function as in (3.36). Define the empirical survival functions for the null part and signal part

$$\widehat{W}_{\text{null}}(t) = \frac{1}{p - s} \sum_{i \notin S} \mathbb{1}\{x(i) \geq t\}, \quad \widehat{W}_{\text{signal}}(t) = \frac{1}{s} \sum_{i \in S} \mathbb{1}\{x(i) \geq t\}, \quad \text{(A.28)}$$

where $s = |S|$, so that

$$\widehat{G}(t) = \frac{p - s}{p} \widehat{W}_{\text{null}}(t) + \frac{s}{p} \widehat{W}_{\text{signal}}(t).$$

Apply Lemma 3.1 to the two summands in $\widehat{G}$, we obtain $\widehat{G}(t) = G(t) + \widehat{R}(t)$. where

$$G(t) = \frac{p - s}{p} \overline{F_0}(t) + \frac{s}{p} \overline{F_a}(t), \quad \text{(A.29)}$$

where $\overline{F_0}$ and $\overline{F_a}$ are the survival functions of $\chi_\nu^2(0)$ and $\chi_\nu^2(\underline{\Delta})$ respectively, and

$$\widehat{R}(t) = O_{\mathbb{P}}\left(\xi_p \sqrt{\overline{F_0}(t) F_0(t)} + \frac{s}{p} \xi_s \sqrt{\overline{F_a}(t) F_a(t)}\right), \quad \text{(A.30)}$$

uniformly in t.

Recall (see proof of Lemma 3.2) that the BH procedure is the thresholding procedure with threshold set at τ (defined in (3.37)). The NDP may also be re-written as

$$\text{NDP} = \frac{|S \setminus \widehat{S}|}{|S|} = \frac{1}{s} \sum_{i \in S} \mathbb{1}\{x(i) < \tau\} = 1 - \widehat{W}_{\text{signal}}(\tau),$$

so that it suffices to show that

$$\widehat{W}_{\text{signal}}(\tau) \to 1 \quad \text{(A.31)}$$

in probability. Applying Lemma 3.1 to $\widehat{W}_{\text{signal}}$, we know that

$$\widehat{W}_{\text{signal}}(\tau) = \overline{F_a}(\tau) + O_{\mathbb{P}}\left(\xi_s \sqrt{\overline{F_a}(\tau) F_a(\tau)}\right) = \overline{F_a}(\tau) + o_{\mathbb{P}}(1).$$

So it suffices to show that $F_a(\tau) \to 0$ in probability. Now let $t^* = 2q \log(p)$ for some q such that $\beta < q < \underline{r}$. We have

$$F_a(t^*) = \mathbb{P}[\chi_\nu^2(\Delta) \le t^*] \le \mathbb{P}\left[2\sqrt{\Delta}Z_\nu \le t^* - \Delta\right]$$

$$= \mathbb{P}\left[Z_\nu \le \frac{t^*}{2\sqrt{\Delta}} - \frac{\sqrt{\Delta}}{2}\right] = \mathbb{P}\left[Z_\nu \le \frac{q-r}{2\sqrt{r}}\sqrt{2\log p}\right] \to 0. \quad \text{(A.32)}$$

Hence in order to show (A.31), it suffices to show

$$\mathbb{P}\left[\tau \le t^*\right] \to 1. \quad \text{(A.33)}$$

By (A.29), the mean of the empirical process $\widehat{G}$ evaluated at t^* is

$$G(t^*) = \frac{p-s}{p}\overline{F_0}(t^*) + \frac{s}{p}\overline{F_a}(t^*). \quad \text{(A.34)}$$

The first term, using Relation (A.24), is asymptotic to $p^{-q}L(p)$, where $L(p)$ is the logarithmic term in p. The second term, since $\overline{F_a}(t^*) \to 1$ by Relation (A.32), is asymptotic to $p^{-\beta}$. Therefore, $G(t^*) \sim p^{-q}L(p) + p^{-\beta} \sim p^{-\beta}$, since $p^{\beta-q}L(p) \to 0$ where $q > \beta$.

The fluctuation of the empirical process at t^*, by Relation (A.30), is

$$\widehat{R}(t^*) = O_{\mathbb{P}}\left(\xi_p\sqrt{\overline{F_0}(t^*)F_0(t^*)} + \frac{s}{p}\xi_s\sqrt{\overline{F_a}(t^*)F_a(t^*)}\right)$$

$$= O_{\mathbb{P}}\left(\xi_p\sqrt{\overline{F_0}(t^*)}\right) + o_{\mathbb{P}}\left(p^{-\beta}\right).$$

By (A.24) and the expression for ξ_p, the first term is $O_{\mathbb{P}}\left(p^{-(q+1)/2}L(p)\right)$ where $L(p)$ is a poly-logarithmic term in p. Since $\beta < \min\{q, 1\}$, we have $\beta < (q+1)/2$, and hence $\widehat{R}(t^*) = o_{\mathbb{P}}(p^{-\beta})$.

Putting the mean and the fluctuation of $\widehat{G}(t^*)$ together, we obtain

$$\widehat{G}(t^*) = G(t^*) + \widehat{R}(t^*) \sim_{\mathbb{P}} G(t^*) \sim p^{-\beta},$$

and therefore, together with (A.24), we have

$$\overline{F_0}(t^*)/\widehat{G}(t^*) = p^{\beta-q}L(p)(1 + o_{\mathbb{P}}(1)),$$

which is eventually smaller than the FDR level α by the assumption (3.17) and the fact that $\beta < q$. That is,

$$\mathbb{P}\left[\overline{F_0}(t^*)/\widehat{G}(t^*) < \alpha\right] \to 1.$$

By definition of τ (recall (3.37)), this implies that $\tau \le t^*$ with probability tending to 1, and (A.33) is shown. The proof for the sufficient condition is complete. $\qquad\square$

A.4 Proof of Theorems 7.2 and 7.4

As with the proof of Theorem 7.3, one could shorten the presentations in this section by referencing arguments in Chap. 3.

Proof (*Theorem 7.2*) We first show the sufficient condition. Similar to the proof of Theorem 7.3, it suffices to show that

$$\text{NDP} = 1 - \widehat{W}_{\text{signal}}(t_p) \to 0, \tag{A.35}$$

where t_p is the threshold of Bonferroni's procedure.

Since $\underline{r} > f_{\text{EA}}(\beta) = 1$, we can pick q such that $1 < q < \underline{r}$. Let $t^* = 2q \log p$, we have $t_p < t_p^*$ for large p as in the proof of Theorem 7.1. Therefore for large p, we have

$$\widehat{W}_{\text{signal}}(t_p) \geq \widehat{W}_{\text{signal}}(t^*) \geq \overline{F_a}(t^*) + o_{\mathbb{P}}(1),$$

where the last inequality follows from the stochastic monotonicity of the chi-square family (Lemma A.2), and Lemma 3.1. Indeed, $F_a(t^*) \to 0$ by (A.32) and our choice of $q < \underline{r}$. The proof of the sufficient condition is complete.

Proof of the necessary condition follows a similar structure to that of Theorem 7.3. That is, we show that FWER + FNR has liminf at least 1 by working with the lower bound

$$\text{FWER}(\mathcal{R}) + \text{FNR}(\mathcal{R}) \geq \mathbb{P}\left[\max_{i \in S^c} x(i) > u\right] \wedge \mathbb{E}\left[\frac{|S \setminus \widehat{S}(u)|}{|S|}\right], \tag{A.36}$$

which holds for any thresholding procedure $\mathcal{R}$ and for arbitrary $u \in \mathbb{R}$. By the assumption that $\bar{r} < f_{\text{EA}}(\beta) = 1$, we can pick q such that $\bar{r} < q < 1$ and let $u = t^* = 2q \log p$. By relative stability of chi-squared random variables (Lemma A.1), we have

$$\mathbb{P}\left[\frac{\max_{i \in S^c} x(i)}{2 \log p} > \frac{t^*}{2 \log p}\right] \to 1. \tag{A.37}$$

where the first fraction in (A.37) converges to 1, while the second converges to $q < 1$. On the other hand, by our choice of $q > \bar{r}$, the second term in (A.36) also converges to 1 as in (A.27). This completes the proof of the necessary condition. $\square$

Proof (*Theorem 7.4*) We first show the sufficient condition. Since FDR control is guaranteed by the BH procedure, we only need to show that the FWNR also vanishes, that is,

$$\mathbb{P}\left[\min_{i \in S} x(i) \geq \tau\right] \to 1, \tag{A.38}$$

where τ is the threshold for the BH procedure.

By the assumption that $\underline{r} > f_{\text{AE}}(\beta) = (\sqrt{\beta} + \sqrt{1-\beta})^2$, we have $\sqrt{\underline{r}} - \sqrt{1-\beta} > \sqrt{\beta}$, so we can pick $q > 0$, such that

$$\sqrt{\underline{r}} - \sqrt{1 - \beta} > \sqrt{q} > \sqrt{\beta}. \tag{A.39}$$

Let $t^* = 2q \log p$, we claim that

$$\mathbb{P}\left[\tau \leq t^*\right] \to 1. \tag{A.40}$$

Indeed, by our choice of $q > \beta$, (A.40) follows in the same way that (A.33) did. With this t^*, we have

$$\mathbb{P}\left[\min_{i \in S} x(i) \geq \tau\right] \geq \mathbb{P}\left[\min_{i \in S} x(i) \geq t^*, \ t^* \geq \tau\right]. \tag{A.41}$$

However, by our choice of $\sqrt{q} < \sqrt{\underline{r}} - \sqrt{1 - \beta}$, the probability of the first event on the right-hand side of (A.41) also goes to 1 according to (A.8) and (A.9). Together with (A.40), this proves (A.38), and completes proof of the sufficient condition.

The necessary condition follows from the lower bound

$$\text{FDR}(\mathcal{R}) + \text{FWNR}(\mathcal{R}) \geq \mathbb{E}\left[\frac{|\widehat{S}(u) \setminus S|}{|\widehat{S}(u) \setminus S| + |S|}\right] \wedge \mathbb{P}\left[\min_{i \in S} x(i) < u\right], \tag{A.42}$$

which holds for any thresholding procedure $\mathcal{R}$ and for arbitrary $u \in \mathbb{R}$.

By the assumption that $\bar{r} < f_{\text{AE}}(\beta) = (\sqrt{\beta} + \sqrt{1 - \beta})^2$, we can pick a constant $q > 0$, such that

$$\sqrt{\bar{r}} - \sqrt{1 - \beta} < \sqrt{q} < \sqrt{\beta}. \tag{A.43}$$

Let also $u = t^* = 2q \log p$. By our choice of $q < \beta$, we know from (A.26) that the first term on the right-hand-side of (A.42) converges to 1. It remains to show that the second term in (A.42) also converges to 1.

For the second term in (A.42), dividing through by $2 \log p$, we obtain

$$\mathbb{P}\left[\min_{i \in S} x(i) < t^*\right] = \mathbb{P}\left[\frac{m_S}{2 \log p} < q\right]. \tag{A.44}$$

Similar to (A.12), we have

$$\frac{m_S}{2 \log p} \overset{d}{\leq} \min_{i \in S} \frac{Z_1^2(i) + \ldots + Z_{\nu-1}^2(i)}{2 \log p} + \frac{(Z_\nu(i) + \sqrt{\overline{\Delta}})^2}{2 \log p}. \tag{A.45}$$

Define $i^\dagger = i_p^\dagger$ to be the index minimizing the second term in (A.45), i.e.,

$$i^\dagger := \arg\min_{i \in S} f_p\left(Z_\nu(i)\right), \tag{A.46}$$

where $f_p(x) := (x + \sqrt{\overline{\Delta}})^2 / (2 \log p)$.

Since $\sqrt{q} > \sqrt{r} - \sqrt{1-\beta}$ and $q > 0$, we have $\frac{\sqrt{r}-\sqrt{q}}{\sqrt{1-\beta}} < 1$. Also, since

$$\frac{\sqrt{r}+\sqrt{q}}{\sqrt{1-\beta}} > 0, \quad \text{and} \quad \frac{\sqrt{r}-\sqrt{q}}{\sqrt{1-\beta}} < \frac{\sqrt{r}+\sqrt{q}}{\sqrt{1-\beta}},$$

we can further pick a constant $\beta_0 \in (0, 1]$ such that

$$\frac{\sqrt{r}-\sqrt{q}}{\sqrt{1-\beta}} < \sqrt{\beta_0} < \frac{\sqrt{r}+\sqrt{q}}{\sqrt{1-\beta}}. \tag{A.47}$$

Let $Z_{[1]} \le Z_{[2]} \le \ldots \le Z_{[s]}$ be the order statistics of $\{Z_\nu(i)\}_{i\in S}$ and define $k = \lfloor s^{1-\beta_0} \rfloor$. Applying Lemma A.4 (stated below), we obtain

$$\frac{Z_{[k]}}{\sqrt{2\log p}} = \frac{Z_{[k]}}{\sqrt{2\log s}}\frac{\sqrt{2\log s}}{\sqrt{2\log p}} \to -\sqrt{\beta_0(1-\beta)} \quad \text{in probability.} \tag{A.48}$$

Since we know (by solving a quadratic inequality) that

$$f_p(x) < q \iff \frac{x}{\sqrt{2\log p}} \in \left(-(\sqrt{r}+\sqrt{q}), -(\sqrt{r}-\sqrt{q})\right), \tag{A.49}$$

combining (A.47), (A.48), and (A.49), it follows that

$$\mathbb{P}\left[f_p\left(Z_\nu(i^\dagger)\right) < q\right] \ge \mathbb{P}\left[f_p\left(Z_{[k]}\right) < q\right] \to 1.$$

Finally, using (A.19), we conclude that

$$\mathbb{P}\left[\min_{i\in S} x(i) < t^*\right] = \mathbb{P}\left[\frac{m_S}{2\log p} < q\right] \ge \mathbb{P}\left[o_\mathbb{P}(1) + f_p\left(Z_\nu(i^\dagger)\right) < q\right] \to 1.$$

Therefore, the two terms on the right-hand-side of (A.42) both converge 1. This completes the proof of the necessary condition. □

It only remains to justify (A.48).

Lemma A.4 (Relative stability of order statistics) *Let $Z_{[1]} \le \ldots \le Z_{[s]}$ be the order statistics of s iid standard Gaussian random variables. Let $\beta_0 \in (0, 1]$ and define $k = \lfloor s^{1-\beta_0} \rfloor$, then we have*

$$\frac{Z_{[k]}}{\sqrt{2\log s}} \to -\sqrt{\beta_0} \quad \text{in probability.} \tag{A.50}$$

Proof *(Lemma A.4)* Using the Renyi representation for order statistics, we write

$$Z_{[i]} = \Phi^{\leftarrow}(U_{[i]}), \tag{A.51}$$

where $U_{[i]}$ is the ith (smallest) order statistic of s independent uniform random variables over $(0, 1)$. Since $U_{[i]}$ has a Beta$(i, s + 1 - i)$ distribution, with mean and standard deviation,

$$\mathbb{E}[U_{[k]}] = k/(s+1) \sim s^{-\beta_0}, \quad \text{and} \quad \text{sd}(U_{[k]}) = \frac{1}{s+1}\sqrt{\frac{k(s+1-k)}{s+2}} \sim s^{-\frac{1+\beta_0}{2}},$$

we obtain by Chebyshev's inequality

$$\mathbb{P}\left[s^{-\beta_0}(1-\epsilon) < U_{[k]} < s^{-\beta_0}(1+\epsilon)\right] \to 1,$$

where ϵ is an arbitrary positive constant. This implies, by representation (A.51),

$$\mathbb{P}\left[\Phi^{\leftarrow}\left(s^{-\beta_0}(1-\epsilon)\right) < Z_{[k]} < \Phi^{\leftarrow}\left(s^{-\beta_0}(1+\epsilon)\right)\right] \to 1. \qquad (A.52)$$

Using the expression for standard Gaussian quantiles (see, e.g., Proposition 1.1. in Gao and Stoev 2020), we know that

$$\Phi^{\leftarrow}\left(s^{-\beta_0}(1-\epsilon)\right) \sim -\sqrt{2\log\left(s^{\beta_0}/(1-\epsilon)\right)}$$
$$= -\sqrt{2(\beta_0 \log s - \log(1-\epsilon))} \sim -\sqrt{2\beta_0 \log s},$$

and similarly $\Phi^{\leftarrow}\left(s^{-\beta_0}(1+\epsilon)\right) \sim -\sqrt{2\beta_0 \log s}$. Since both ends of the interval in (A.52) are asymptotic to $-\sqrt{2\beta_0 \log s}$, the desired conclusion follows. $\qquad \square$

Appendix B
Exact Support Recovery in Non AGG Models

B.1 Strong Classification Boundaries in Other Light-Tailed Error Models

The strong classification boundaries extend beyond the AGG models. As our analysis in Chap. 4 suggests, all additive error models where the errors have URS maxima exhibit this phase transition phenomenon under appropriate parametrization of the sparsity and signal sizes. We derive explicit boundaries for two additional classes of models under the general form of the additive noise models (1.1) with *heavier* and *lighter* tails than the AGG models, respectively.

We would like to point out that the sparsity and signal sizes can be re-parametrized for the boundaries to have different shapes. For example in the case of Gaussian errors, if we re-parametrize sparsity s with $\widetilde{\beta} = 2 - \left(1 + \sqrt{1-\beta}\right)^2$ where $\widetilde{\beta} \in (0, 1)$, then the signal sparsity would have a slightly more complicated form:

$$|S_p| = \lfloor p^{1-\beta} \rfloor = \left\lfloor p^{\left(\sqrt{2-\widetilde{\beta}}-1\right)^2} \right\rfloor,$$

while the strong classification boundary would take on the simpler form:

$$f_E(\beta) = \widetilde{f}_E(\widetilde{\beta}) = 2 - \widetilde{\beta}. \tag{B.1}$$

In the next two classes of models we will adopt parametrizations such that the boundaries are of the form $\widetilde{g}$ in (B.1).

© The Author(s), under exclusive license to Springer Nature Switzerland AG 2021
Z. Gao and S. Stoev, *Concentration of Maxima and Fundamental
Limits in High-Dimensional Testing and Inference*,
SpringerBriefs in Probability and Mathematical Statistics,
https://doi.org/10.1007/978-3-030-80964-5

B.1.1 Additive Error Models with Heavier-Than-AGG Tails

Distributions such as the log-normal have heavier tails than the AGG model, yet the tails are nevertheless rapidly-varying. Therefore, Proposition 2.2 applies, and we expect to see phase-transition-type results when the additive errors have these heavier-than-AGG tails.

Example B.1 (*Heavier than AGG*) Let $\gamma > 1, c > 0$, and suppose that

$$\log \overline{F}(x) = -(\log x)^{\gamma} (c + M(x)), \tag{B.2}$$

where $\lim_{x \to \infty} M(x) \log^{\gamma} x = 0$. Then, Relation (2.39) holds under model (B.2). Further, if the entries in the array are independent, the maxima are relatively stable.

The behavior of the quantiles u_p in this model is as follows. As $p \to \infty$,

$$u_p \sim \exp \left\{ \left(c^{-1} \log p \right)^{1/\gamma} \right\} \iff c \left(\log u_p \right)^{\gamma} + o(1) = \log(p) = -\log \overline{F}(u_p).$$

since u_p diverges, and $M(u_p)$ is $o((\log^{\gamma} u_p)^{-1})$.

Following Example B.1, assume that the errors in Model (1.1) have rapidly varying right tails

$$\log \overline{F}(x) = -(\log x)^{\gamma} (c + M(x)), \tag{B.3}$$

as $x \to \infty$, and left tails

$$\log F(x) = -(\log(-x))^{\gamma} (c + M(-x)), \tag{B.4}$$

as $x \to -\infty$.

Theorem B.1 *Suppose the marginals F follows* (B.3) *and* (B.4). *Let*

$$k(\beta) = \log p - \left((\log p)^{1/\gamma} + \log (1 - \beta) \right)^{\gamma},$$

and let the signal μ have

$$|S_p| = \left\lfloor p e^{-k(\beta)} \right\rfloor$$

non-zero entries. Assume the magnitudes of non-zero signal entries are in the range between

$$\underline{\Delta} = \exp \left\{ (\log p)^{1/\gamma} \right\}^{\underline{r}} \quad and \quad \overline{\Delta} = \exp \left\{ (\log p)^{1/\gamma} \right\}^{\overline{r}}.$$

If $\underline{r} > \widetilde{f_E}(\beta) = 2 - \beta$, then Bonferroni's procedure $\widehat{S}_p$ (defined in (2.21)) with appropriately calibrated FWER $\alpha \to 0$ achieves asymptotic perfect support recovery, under arbitrary dependence of the errors.

On the other hand, when the errors are uniformly relatively stable, if $\overline{r} < \widetilde{f_E}(\beta) = 2 - \beta$, then no thresholding procedure can achieve asymptotic perfect support recovery with positive probability.

B.1.2 Additive Error Models with Lighter-Than-AGG Tails

Similar to how Proposition 2.2 applies to models with heavier-than-AGG tails, it also to error models with lighter tails than the AGG class.

Example B.2 (*Lighter than AGG*) With $\nu > 0$, and $L(x)$ a slowly varying function, the class of distributions

$$\log \overline{F}(x) = -\exp\{x^\nu L(x)\}, \tag{B.5}$$

is rapidly varying. The quantiles can be derived explicitly in a subclass of (B.5) where $L(x) \to 1$, or equivalently, when $\log|\log \overline{F}(x)| \sim x^\nu$,

$$u_p \sim (\log\log p)^{1/\nu} \iff \exp\{u_p^\nu (1 + o(1))\} = \log(p) = -\log \overline{F}(u_p).$$

Following Example B.2, assume that errors in Model (1.1) has rapidly varying right tails

$$\log \overline{F}(x) = -\exp\{x^\nu L(x)\}, \tag{B.6}$$

where $L(x)$ is a slowly varying function, as $x \to \infty$, and left tails

$$\log \overline{F}(x) = -\exp\{-x^\nu L(-x)\}, \tag{B.7}$$

as $x \to -\infty$.

The phase transition results in multiple testing problems under such tail assumptions is characterizes as follows.

Theorem B.2 *Suppose marginals F follow (B.6) and (B.7). Let*

$$k(\beta) = \log p - (\log(p))^{(1-\beta)^\nu},$$

and let the signal μ have

$$|S_p| = \left\lfloor p e^{-k(\beta)} \right\rfloor$$

non-zero entries. Assume the magnitudes of non-zero signal entries are in the range between

$$\underline{\Delta} = \log\log p^{1/\nu}\underline{r} \quad and \quad \overline{\Delta} = \log\log p^{1/\nu}\overline{r}.$$

If $\underline{r} > \widetilde{f}_E(\beta) = 2 - \beta$, then Bonferroni's procedure $\widehat{S}_p$ (defined in (2.21)) with appropriately calibrated FWER $\alpha \to 0$ achieves asymptotic perfect support recovery, under arbitrary dependence of the errors.

On the other hand, when the errors are uniformly relatively stable, if $\overline{r} < \widetilde{f}_E(\beta) = 2 - \beta$, then no thresholding procedure can achieve asymptotic perfect support recovery with positive probability.

B.2   Thresholding Procedures Under Heavy-Tailed Errors

We analyze the performance of thresholding estimators under heavy-tailed models in this section, and illustrate its lack of phase transition. Suppose we have iid errors with Pareto tails in Model (1.1), that is, $\epsilon(i)$'s have common marginal distribution F where

$$\overline{F}(x) \sim x^{-\alpha} \quad \text{and} \quad F(-x) \sim x^{-\alpha}, \tag{B.8}$$

as $x \to \infty$. It is well-known (see, e.g., Theorem 1.6.2 of Leadbetter et al. 1983) that the maxima of iid Pareto random variables have Frechet-type limits. Specifically, we have

$$\frac{\max_{i \in \{1,\dots,p\}} \epsilon(i)}{u_p} \implies Y, \tag{B.9}$$

in distribution, where $u_p = F^{\leftarrow}(1 - 1/p) \sim p^{1/\alpha}$, and Y is a standard α-Frechet random variable, i.e.,

$$\mathbb{P}[Y \le t] = \exp\{-t^{-\alpha}\}, \quad t > 0.$$

By symmetry in our assumptions, the same argument applies to the minima as well.

Theorem B.3 *Let errors in Model (1.1) be as described in Relation (B.8). Let the signal have $s = |S| = fp$ non-zero entries, with magnitude $\Delta = rp^{1/\alpha}$, where both $f \in (0, 1)$ and $r \in (0, +\infty)$ may depend on p, so that no generality is lost.*

Under these assumptions, the necessary condition for thresholding procedures $\widehat{S}$ to achieve exact support recovery ($\mathbb{P}[\widehat{S} = S] \to 1$) is

$$\liminf_{p \to \infty} r = \infty. \tag{B.10}$$

Condition (B.10) is also sufficient for the oracle thresholding procedure to succeed in the exact support recovery problem.

On the other hand, the necessary and sufficient condition for all thresholding procedures to fail exact support recovery ($\mathbb{P}[\widehat{S} = S] \to 0$) is

$$\limsup_{p \to \infty} r = 0.$$

In other words, Theorem B.3 states that there does not exist a non-trivial phase transition for thresholding procedures when errors have (two-sided) α-Pareto tails.

Proof (*Theorem* B.3) Recall the oracle thresholding procedure $\widehat{S}^* = \{i : x(i) \ge x_{[s]}\}$, and the set of all thresholding procedures, denoted $\mathcal{S}$ (see Definition 2.20). The probability of exact support recovery by any thresholding procedure $\widehat{S} \in \mathcal{S}$ is bounded above by that of $\widehat{S}^*$, that is,

$$\max_{\widehat{S} \in \mathcal{S}} \mathbb{P}[\widehat{S} = S] = \mathbb{P}[\widehat{S}^* = S] = \mathbb{P}\left[\max_{i \in S^c} x(i) \leq \min_{i \in S} x(i) \right]$$

$$= \mathbb{P}\left[\frac{\max_{i \in S^c} x(i)}{u_p} \leq \frac{\min_{i \in S} x(i)}{u_p} \right]$$

$$= \mathbb{P}\left[\frac{M_{S^c}}{u_p} \leq \frac{m_S}{u_p} + r_p \right], \qquad \text{(B.11)}$$

where $M_{S^c} = \max_{i \in S^c} \epsilon(i)$ and $m_S = \min_{i \in S} \epsilon(i)$. For any $\alpha > 0$, the following elementary relations hold,

$$0 < L \leq (1 - f)^{1/\alpha} + f^{1/\alpha} \leq U < \infty, \quad \text{for all } f \in (0, 1),$$

where $L = \min\left\{1, 2(1/2)^{1/\alpha}\right\}$ and $U = \max\left\{1, 2(1/2)^{1/\alpha}\right\}$. Therefore we have,

$$U \max\left\{ \frac{M_{S^c}}{u_p}, -\frac{m_S}{u_p} \right\} < r_p \implies (1 - f)^{1/\alpha} \frac{M_{S^c}}{u_p} - f^{1/\alpha} \frac{m_S}{u_p} < r_p, \qquad \text{(B.12)}$$

and

$$L \min\left\{ \frac{M_{S^c}}{u_p}, -\frac{m_S}{u_p} \right\} < r_p \impliedby (1 - f)^{1/\alpha} \frac{M_{S^c}}{u_p} - f^{1/\alpha} \frac{m_S}{u_p} < r_p. \qquad \text{(B.13)}$$

Putting together (B.11), (B.12), and (B.13), we have

$$\mathbb{P}\left[\max\left\{ \frac{M_{S^c}}{u_p}, -\frac{m_S}{u_p} \right\} < r_p/U \right] \leq \mathbb{P}[\widehat{S}^* = S] \leq \mathbb{P}\left[\min\left\{ \frac{M_{S^c}}{u_p}, -\frac{m_S}{u_p} \right\} < r_p/L \right].$$
$$\text{(B.14)}$$

We know from the weak convergence result (B.9) that for any $\epsilon > 0$ there is a constant N such that for all $p > N$ we have

$$\mathbb{P}\left[\max\left\{ \frac{M_{S^c}}{u_p}, -\frac{m_S}{u_p} \right\} < r_p/U \right] \geq \mathbb{P}\left[\max\left\{ Y^{(1)}, Y^{(2)} \right\} < r_p/U \right] - \epsilon, \qquad \text{(B.15)}$$

where $Y^{(1)}$ and $Y^{(2)}$ are independent α-Frechet random variables with scale coefficients $(1 - f)^{1/\alpha}$ and $f^{1/\alpha}$ respectively. That is,

$$\mathbb{P}[Y^{(1)} \leq t] = \exp\left\{-(1 - f)/t^\alpha\right\}, \quad \text{and} \quad \mathbb{P}[Y^{(2)} \leq t] = \exp\left\{-f/t^\alpha\right\}.$$

Since the distributional limit in (B.15) has a density (with respect to the Lebesgue measure), we know that density is bounded above by a finite constant, say, K. For the same choice of ϵ as before, we can find a further constant N' such that for all $p > \max\{N, N'\}$ we have

$$\liminf r_p < \epsilon/K + r_p,$$

so that the right hand side of (B.15) is bounded by

$$\mathbb{P}\left[\max\left\{Y^{(1)}, Y^{(2)}\right\} < r_p/U\right] - \epsilon \geq \mathbb{P}\left[\max\left\{Y^{(1)}, Y^{(2)}\right\} < \frac{\liminf r_p}{U}\right] - 2\epsilon.$$
(B.16)

By the arbitrariness in the choice of ϵ, we conclude from (B.15) and (B.16) that

$$\liminf \mathbb{P}\left[\max\left\{\frac{M_{S^c}}{u_p}, -\frac{m_S}{u_p}\right\} < r_p/U\right] \geq \mathbb{P}\left[\max\left\{Y^{(1)}, Y^{(2)}\right\} < \frac{\liminf r_p}{U}\right].$$
(B.17)

Combining Relations (B.14) and (B.17), we know that if $\liminf r_p = \infty$, we must have

$$\liminf \mathbb{P}\left[\widehat{S}^* = S\right] \geq \mathbb{P}\left[\max\left\{Y^{(1)}, Y^{(2)}\right\} < \frac{\liminf r_p}{U}\right] = 1.$$

Conversely, if $\liminf \mathbb{P}\left[\widehat{S}^* = S\right] < 1$, we must have $\liminf r_p < \infty$.

Similarly, we can obtain the upper bound of exact support recovery probability for the optimal thresholding procedure,

$$\limsup \mathbb{P}\left[\min\left\{\frac{M_{S^c}}{u_p}, -\frac{m_S}{u_p}\right\} < r_p/L\right] \leq \mathbb{P}\left[\min\left\{Y^{(1)}, Y^{(2)}\right\} < \frac{\limsup r_p}{L}\right].$$
(B.18)

The conclusions of the second part of Theorem B.3 follow from (B.14) and (B.18). $\square$

The probability of exact recovery can be approximated if the parameters r and f converge. The next result follows from a small modification of the arguments in the proof of Theorem B.3.

Corollary B.1 *Under the assumptions in Theorem B.3, if* $\lim r = r^*$, *and* $\lim f = f^*$, *for some constant* $r^* \geq 0$ *and* $f^* \in [0, 1]$, *then*

$$\lim \mathbb{P}[\widehat{S}^* = S] = \mathbb{P}\left[(1 - f^*)^{1/\alpha} Z_1 + (f^*)^{1/\alpha} Z_2 < r^*\right].$$

where Z_1 and Z_2 are independent standard α-Frechet random variables, i.e., $\mathbb{P}[Z_i \leq x] = \exp\{-x^{-\alpha}\}, x > 0$.

Remark B.1 Of course one might wonder if it would be meaningful to derive a "phase transition" under a different parametrization of the signal sizes, say

$$\Delta = p^{r/\alpha}.$$
(B.19)

In this case, Theorem B.3 suggests that a "phase transition" takes place at $r = 1$. However, this non-multiplicative parametrization of the signal sizes would make power analysis (like in Example 3.1) dimension-dependent.

To illustrate, in the case of Gaussian errors with variance 1, if we were interested in small signals of size $\sqrt{2r \log p}$, where $r < 1$ is below the boundary (4.5), then we only need $n > 2/r$ samples to guarantee discovery of their support. In the Pareto case with parametrization (B.19), however, if we were interested in small signals of size $p^{r/\alpha}$, where $r < 1$, then the "boundary" says that we will need $n > p^{2(1-r)/\alpha}$ samples, which is exponential in the dimension p and quickly diverges. Recall that the "boundary" is really an asymptotic result in p. Such an approximation in finite dimensions becomes invalid.

Bibliography

Adler, R.J., Taylor, J.E.: Random Fields and Geometry. Springer Science & Business Media, Berlin (2009)

Agresti, A.: An Introduction to Categorical Data Analysis. Wiley, Hoboken (2018)

Anderson, T.W., Darling, D.A.: Asymptotic theory of certain "goodness of fit" criteria based on stochastic processes. Ann. Math. Stat. 193–212 (1952)

Arias-Castro, E., Chen, S.: Distribution-free multiple testing. Electron. J. Stat. **11**(1), 1983–2001 (2017)

Arias-Castro, E., Wang, M.: Distribution-free tests for sparse heterogeneous mixtures. Test **26**(1), 71–94 (2017)

Barber, R.F., Candès, E.J.: Controlling the false discovery rate via knockoffs. Ann. Stat. **43**(5), 2055–2085 (2015)

Barndorff-Nielsen, O.: On the limit behaviour of extreme order statistics. Ann. Math. Stat. **34**(3), 992–1002 (1963)

Benjamini, Y., Hochberg, Y.: Controlling the false discovery rate: a practical and powerful approach to multiple testing. J. R. Stat. Society. Ser. B (Methodol.) 289–300 (1995)

Benjamini, Y., Yekutieli, D.: The control of the false discovery rate in multiple testing under dependency. Ann. Stat. 1165–1188 (2001)

Berman, S.M.: Limit theorems for the maximum term in stationary sequences. Ann. Math. Stat. 502–516 (1964)

Bingham, N.H., Goldie, C.M., Teugels, J.L.: Regular Variation. Cambridge University Press, Cambridge (1987)

Bogdan, M., Chakrabarti, A., Frommlet, F., Ghosh, J.K.: Asymptotic Bayes-optimality under sparsity of some multiple testing procedures. Ann. Stat. **39**(3), 1551–1579 (2011)

Boucheron, S., Lugosi, G., Massart, P.: Concentration Inequalities. A Nonasymptotic Theory of Independence, With a foreword by Michel Ledoux. Oxford University Press, Oxford (2013)

Bush, W.S., Moore, J.H.: Genome-wide association studies. PLoS Comput. Biol. **8**(12), e1002822 (2012)

Butucea, C., Ndaoud, M., Stepanova, N.A., Tsybakov, A.B.: Variable selection with Hamming loss. Ann. Stat. **46**(5), 1837–1875 (2018)

Cai, T.T., Jeng, X.J., Jin, J.: Optimal detection of heterogeneous and heteroscedastic mixtures. J. R. Stat. Soc.: Ser. B (Stat. Methodol.) **73**(5), 629–662 (2011)

© The Author(s), under exclusive license to Springer Nature Switzerland AG 2021

Z. Gao and S. Stoev, *Concentration of Maxima and Fundamental Limits in High-Dimensional Testing and Inference*, SpringerBriefs in Probability and Mathematical Statistics, https://doi.org/10.1007/978-3-030-80964-5

Cai, T.T., Jin, J., Low, M.G.: Estimation and confidence sets for sparse normal mixtures. Ann. Stat. **35**(6), 2421–2449 (2007)

Cai, T.T., Wu, Y.: Optimal detection of sparse mixtures against a given null distribution. IEEE Trans. Inf. Theory **60**(4), 2217–2232 (2014)

Candès, E.J.: Lecture 3: global testing, chi-square test, optimality of chi-square test for distributed mild effects. In: Stats 300C: Theory of Statistics (Spring 2018). Stanford Lecture Notes (2018). https://statweb.stanford.edu/~candes/teaching/stats300c/index.html

Chatterjee, S.: Superconcentration and Related Topics, vol. 15. Springer, Berlin (2014)

Comminges, L., Dalalyan, A.S.: Tight conditions for consistency of variable selection in the context of high dimensionality. Ann. Stat. **40**(5), 2667–2696 (2012)

Conlon, D., Fox, J., Sudakov, B.: Recent developments in graph Ramsey theory. In: Surveys in Combinatorics 2015. London Mathematical Society Lecture Note Series, vol. 424, pp. 49–118. Cambridge University Press, Cambridge (2015)

Cramér, H.: On the composition of elementary errors: first paper: mathematical deductions. Scand. Actuar. J. **1928**(1), 13–74 (1928)

De Haan, L., Ferreira, A.: Extreme Value Theory: An Introduction. Springer Science & Business Media, Berlin (2007)

Dedecker, J., Doukhan, P., Lang, G., León, J.R., Louhichi, S., Prieur, C.: Weak Dependence: With Examples and Applications. Lecture Notes in Statistics, vol. 190. Springer, New York (2007)

Domingos, P., Pazzani, M.: On the optimality of the simple Bayesian classifier under zero-one loss. Mach. Learn. **29**(2–3), 103–130 (1997)

Donoghue, W.F.: Distributions and Fourier Transforms, vol. 32. Academic, Cambridge (2014)

Donoho, D., Jin, J.: Higher criticism for detecting sparse heterogeneous mixtures. Ann. Stat. **32**(3), 962–994 (2004)

Donoho, D., Jin, J.: Special invited paper: higher criticism for large-scale inference, especially for rare and weak effects. Stat. Sci. 1–25 (2015)

Dudoit, S., Shaffer, J.P., Boldrick, J.C.: Multiple hypothesis testing in microarray experiments. Stat. Sci. 71–103 (2003)

Dunn, O.J.: Multiple comparisons among means. J. Am. Stat. Assoc. **56**(293), 52–64 (1961)

Efron, B.: Large-scale simultaneous hypothesis testing: the choice of a null hypothesis. J. Am. Stat. Assoc. **99**(465), 96–104 (2004)

Eicker, F.: The asymptotic distribution of the suprema of the standardized empirical processes. Ann. Stat. 116–138 (1979)

Embrechts, P., Klüppelberg, C., Mikosch, T.: Modelling Extremal Events: For Insurance and Finance, vol. 33. Springer Science & Business Media, Berlin (2013)

Erdös, P., Szekeres, G.: A combinatorial problem in geometry. Compos. Math. **2**, 463–470 (1935)

Fan, J.: Test of significance based on wavelet thresholding and Neyman's truncation. J. Am. Stat. Assoc. **91**(434), 674–688 (1996)

Ferguson, T.S.: A Course in Large Sample Theory. Routledge, Milton Park (2017)

Feynman, R.: The Character of Physical Law. MIT Press, Cambridge (2017)

Fox, J.: Lecture 5: Ramsey Theory. In: MAT 307: Combinatorics (Spring 2009). MIT Lecture Notes (2009). http://math.mit.edu/~fox/MAT307.html

Gao, Z. (2019). U-PASS: https://power.stat.lsa.umich.edu/u-pass/. An R Shiny App accompanying the paper "U-PASS: unified power analysis and forensics for qualitative traits in genetic association studies"

Gao, Z., Stoev, S.: Fundamental limits of exact support recovery in high dimensions. Bernoulli **26**(4), 2605–2638 (2020)

Gao, Z., Terhorst, J., Van Hout, C.V., Stoev, S.: U-PASS: unified power analysis and forensics for qualitative traits in genetic association studies. Bioinformatics **36**(3), 974–975 (2019)

Genovese, C., Wasserman, L.: Operating characteristics and extensions of the false discovery rate procedure. J. R. Stat. Soc.: Ser. B (Stat. Methodol.) **64**(3), 499–517 (2002)

Genovese, C.R., Jin, J., Wasserman, L., Yao, Z.: A comparison of the lasso and marginal regression. J. Mach. Learn. Res. **13**, 2107–2143 (2012)

Gnedenko, B.: Sur la distribution limite du terme maximum d'une serie aleatoire. Ann. Math. 423–453 (1943)

Hall, P., Jin, J.: Innovated higher criticism for detecting sparse signals in correlated noise. Ann. Stat. **38**(3), 1686–1732 (2010)

He, Y., Xu, G., Wu, C., Pan, W.: Asymptotically independent U-statistics in high-dimensional testing (2018). arXiv preprint arXiv:1809.00411

Hochberg, Y.: A sharper Bonferroni procedure for multiple tests of significance. Biometrika **75**(4), 800–802 (1988)

Holm, S.: A simple sequentially rejective multiple test procedure. Scand. J. Stat. 65–70 (1979)

Hsing, T.: A note on the asymptotic independence of the sum and maximum of strongly mixing stationary random variables. Ann. Probab. 938–947 (1995)

Inglot, T.: Inequalities for quantiles of the chi-square distribution. Probab. Math. Stat. **30**(4), 339–351 (2010)

Ingster, Y.I.: Minimax detection of a signal for ℓ_n^p-balls. Math. Methods Stat. **7**(4), 401–428 (1998)

Ji, P., Jin, J.: UPS delivers optimal phase diagram in high-dimensional variable selection. Ann. Stat. **40**(1), 73–103 (2012)

Jin, J., Zhang, C.-H., Zhang, Q.: Optimality of graphlet screening in high dimensional variable selection. J. Mach. Learn. Res. **15**(1), 2723–2772 (2014)

Kallitsis, M., Stoev, S.A., Bhattacharya, S., Michailidis, G.: AMON: an open source architecture for online monitoring, statistical analysis, and forensics of multi-gigabit streams. IEEE J. Sel. Areas Commun. **34**(6), 1834–1848 (2016)

Kartsioukas, R., Gao, Z., Stoev, S.: On the rate of concentration of maxima in Gaussian arrays (2019). arXiv preprint arXiv:1910.04259

Klass, M.J.: The minimal growth rate of partial maxima. Ann. Probab. **12**(2), 380–389 (1984)

Leadbetter, M.R., Lindgren, G., Rootzén, H.: Extremes and Related Properties of Random Sequences and Processes. Springer Series in Statistics, Springer, Berlin (1983)

Li, X., Fithian, W.: Optimality of the max test for detecting sparse signals with Gaussian or heavier tail (2020). arXiv:2006.12489

MacArthur, J., Bowler, E., Cerezo, M., Gil, L., Hall, P., Hastings, E., Junkins, H., McMahon, A., Milano, A., Morales, J.: The new NHGRI-EBI catalog of published genome-wide association studies (GWAS catalog). Nucl. Acids Res. **45**(D1), D896–D901 (2016)

McCormick, W., Mittal, Y.: On weak convergence of the maximum. Department of Statistics, Stanford University (1976)

Meinshausen, N., Bühlmann, P.: High-dimensional graphs and variable selection with the lasso. Ann. Stat. **34**(3), 1436–1462 (2006)

Michailidou, K., Lindström, S., Dennis, J., Beesley, J., Hui, S., Kar, S., Lemaçon, A., Soucy, P., Glubb, D., Rostamianfar, A.: Association analysis identifies 65 new breast cancer risk loci. Nature **551**(7678), 92 (2017)

Naveau, P.: Almost sure relative stability of the maximum of a stationary sequence. Adv. Appl. Probab. **35**(3), 721–736 (2003)

Neuvial, P., Roquain, E.: On false discovery rate thresholding for classification under sparsity. Ann. Stat. **40**(5), 2572–2600 (2012)

Nichols, T., Hayasaka, S.: Controlling the familywise error rate in functional neuroimaging: a comparative review. Stat. Methods Med. Res. **12**(5), 419–446 (2003)

Pipiras, V., Taqqu, M.S.: Long-Range Dependence and Self-similarity. Cambridge Series in Statistical and Probabilistic Mathematics, [45]. Cambridge University Press, Cambridge (2017)

Ramsey, F.P.: On a problem of formal logic. In: Classic Papers in Combinatorics, pp. 1–24. Springer, Berlin (2009)

Resnick, S.I.: Extreme Values, Regular Variation and Point Processes. Springer, Berlin (2013)

Resnick, S.I.: A probability path. Modern Birkhäuser Classics. Birkhäuser/Springer, New York (2014). Reprint of the fifth (2005) printing of the 1999 original [MR1664717]

Resnick, S.I., Tomkins, R.: Almost sure stability of maxima. J. Appl. Probab. **10**(2), 387–401 (1973)

Robbins, H.: A remark on Stirling's formula. Am. Math. Mon. **62**(1), 26–29 (1955)

Sah, A.: Diagonal Ramsey via effective quasirandomness (2020)

Šidák, Z.: Rectangular confidence regions for the means of multivariate normal distributions. J. Am. Stat. Assoc. **62**(318), 626–633 (1967)

Skorokhod, A.V.: Limit theorems for stochastic processes. Theory Probab. Its Appl. **1**(3), 261–290 (1956)

Slepian, D.: The one-sided barrier problem for Gaussian noise. Bell Labs Tech. J. **41**(2), 463–501 (1962)

Smirnov, N.: Table for estimating the goodness of fit of empirical distributions. Ann. Math. Stat. **19**(2), 279–281 (1948)

Storey, J.D.: The optimal discovery procedure: a new approach to simultaneous significance testing. J. R. Stat. Soc.: Ser. B (Stat. Methodol.) **69**(3), 347–368 (2007)

Sun, W., Cai, T.T.: Oracle and adaptive compound decision rules for false discovery rate control. J. Am. Stat. Assoc. **102**(479), 901–912 (2007)

Tanguy, K.: Some superconcentration inequalities for extrema of stationary Gaussian processes. Stat. Probab. Lett. **106**, 239–246 (2015a)

Tanguy, K.: Some superconcentration inequalities for extrema of stationary Gaussian processes. Stat. Probab. Lett. **106**, 239–246 (2015b)

Taqqu, M.S.: Fractional Brownian motion and long-range dependence. In: Doukhan, P., Oppenheim, G., Taqqu, M.S. (eds.) Theory and Applications of Long-Range Dependence, pp. 5–38. Birkhäuser, Basel (2003)

Tsagris, M., Beneki, C., Hassani, H.: On the folded normal distribution. Mathematics **2**(1), 12–28 (2014)

Tukey, J.W.: T13N: the higher criticism. In: Statistics 411. Princeton University Lecture Notes (1976)

Wainwright, M.J.: Information-theoretic limits on sparsity recovery in the high-dimensional and noisy setting. IEEE Trans. Inf. Theory **55**(12), 5728–5741 (2009a)

Wainwright, M.J.: Sharp thresholds for high-dimensional and noisy sparsity recovery using ℓ_1-constrained quadratic programming (Lasso). IEEE Trans. Inf. Theory **55**(5), 2183–2202 (2009b)

Wainwright, M.J.: High-Dimensional Statistics: A Non-asymptotic Viewpoint. Cambridge University Press, Cambridge (2019)

Wasserman, L., Roeder, K.: High dimensional variable selection. Ann. Stat. **37**(5A), 2178 (2009)

Wu, C., Xu, G., Pan, W.: An adaptive test on high-dimensional parameters in generalized linear models. Stat. Sin. **29**, 2163–2186 (2019)

Xu, G., Lin, L., Wei, P., Pan, W.: An adaptive two-sample test for high-dimensional means. Biometrika **103**(3), 609–624 (2016)

Zhang, J.: Powerful goodness-of-fit tests based on the likelihood ratio. J. R. Stat. Soc.: Ser. B (Stat. Methodol.) **64**(2), 281–294 (2002)

Zhao, P., Yu, B.: On model selection consistency of Lasso. J. Mach. Learn. Res. **7**, 2541–2563 (2006)

Zhong, P.-S., Chen, S.X., Xu, M.: Tests alternative to higher criticism for high-dimensional means under sparsity and column-wise dependence. Ann. Stat. **41**(6), 2820–2851 (2013)

Printed in the United States
by Baker & Taylor Publisher Services